P9-EET-993

Clearing the Air

ON November 4–6, 1987, eighty scientists from Europe, Asia, and North America met at the International Conference on Indoor Air Quality in Tokyo, Japan. The conference was hosted by Japan's Council for Environment and Health. At the conclusion of the conference, the organizing committee issued a press release emphasizing the need for further research on the issues of Environmental Tobacco Smoke (ETS) exposures and possible health effects. They noted, specifically, the necessity of giving "priority to solving other public health problems," given the "low probability of proving" the existence of a relationship between ETS and health effects. They concluded that research on various factors in our environment, "not just ETS or passive smoking, must be promoted."

Summaries of other research evidence on ETS appear in the appendix, pages 119–44.

Clearing the Air

Perspectives on Environmental Tobacco Smoke

Edited by

ROBERT D. TOLLISON

Lexington Books

D.C. Heath and Company • Lexington, Massachusetts • Toronto

Library of Congress Cataloging-in-Publication Data

Clearing the air.

1. Air—Pollution, Indoor. 2. Tobacco—Environmental aspects.
3. Tobacco—Toxicology. 4. Smoking—Environmental aspects.
I. Tollison, Robert D.
TD883.1.C58 1988 363.4 87-32513
ISBN 0-669-18007-6 (alk. paper)

Published simultaneously in Canada
Printed in the United States of America
International Standard Book Number: 0-669-18007-6
Library of Congress Catalog Card Number: 87-32513

The paper used in this publication meets the minimum requirements of American National Standard for Information Sciences—Permanence of Paper for Printed Library Materials, ANSI Z39.48-1984.

88 89 90 91 92 8 7 6 5 4 3 2

Contents

Preface

It has become increasingly difficult for the tobacco industry to articulate its views of the appropriate public policies with respect to its products and customers. More often than not, these views are shouted down by antitobacco hotheads who have long since (if they ever had) lost any desire to debate the issues. Yet the tobacco industry has important and useful things to say. Excise taxes are bad taxes; public policy toward smoking does pick on poor people; smoking is like a lot of other risky activities that individuals freely and knowingly undertake but as yet do not seem to raise similar concerns of public policy; and so on. The point is how to get these views into the public arena.

This book is about environmental tobacco smoke, and it has come to fruition with the support of Philip Morris, Inc. This support was useful to our endeavors but in no way influenced the opinions expressed by the authors. Any argument or idea worth its salt has a life of its own. The ideas in this book will withstand critical scrutiny should anyone care to debate them.

1

Introduction and Overview

Robert D. Tollison

ENVIRONMENTAL tobacco smoke (ETS) raises many complex issues for debate and analysis. There are issues concerning possible health effects of ETS on nonsmokers, the measurement of ETS and other substances in indoor environments, economic incentives and the appropriate public policy toward ETS, law and civil liberties, collective bargaining and corporate management, the behavior of interest groups and bureaucracies, and how smoking and ETS differ in a meaningful way from other behavior to which some people may object. This short list undoubtedly does not cover all the potentially relevant issues with respect to ETS.

Unfortunately the complexity of these issues is rarely addressed in public discussions of ETS. Many of the relevant points are made by prosmoking and antismoking advocates, but each side in the debate is presumed to operate strictly in terms of vested interests. The vested interest of the tobacco industry is obvious; the vested interest of the antismoking movement is less obvious but nonetheless very real, consisting of such factors as the size and staff of public health organizations and bureaucracies. Moreover, the tone of the debate about ETS is often shrill and confusing. The two sides seem to talk past one another, seeking primarily to influence public opinion.

The general public is caught in the middle of this debate, and under the circumstances it must be difficult for the average citizen to assess the issues surrounding ETS. If anything, the public may have been influenced by the antismoking movement to believe that ETS is equivalent to a modern plague that kills and harms thousands of people yearly. This is not the case. The basic facts about ETS are quite different from the stereotypical "facts" that have emerged from the public debate thus far.

The purpose of this book is to bring the full range of evidence and points of view about ETS to the surface, where they can be examined and critiqued dispassionately. This end is pursued in two basic ways. In the Appendix, the literature on the potential health consequences of ETS is reviewed in order to highlight the basic conclusions of this research. The quotations contained in the Appendix show that, there is no compelling evidence that normal levels of ETS in indoor environments are a hazard to health and that more and better research designs may prove this basic conclusion to be wrong, but this does not seem very likely. The quotations are based on an extensive review of the scientific literature about ETS and are not taken out of context. To the contrary, they represent the basic conclusions of each scientific article. The Appendix serves as crucial background for the rest of the book.

The main text of the book offers a variety of perspectives on ETS by distinguished individuals. This format is designed to allow readers to obtain a brief overview of the relevant issues in a particular aspect of the debate about ETS. Perspectives range from those of health scientists to those of an economist, a sociologist, a labor union official, a corporate president, and others. In effect, the book is designed to present the ETS issue from several different and important points of view. How does a labor union official evaluate the ETS issue? What is the view of a corporate executive? What are the various scholarly points of view? Perhaps this approach will allow some balance and perspective to be brought to bear on public thinking about ETS. This is the spirit in which the book was designed.

The first two chapters address basic health issues. Mark J. Reasor of the West Virginia University Medical Center offers an overview of the state of the medical and scientific evidence about the potential health effects of ETS. He finds what other scientists have discovered (see Appendix): that there is no convincing evidence that ETS poses any deleterious health problems in normal, indoor environments. W. Allan Crawford, a public health consultant from Australia, reaches similar conclusions with respect to the need for sweeping public health legislation to control ETS. He finds that such laws are not necessitated by the evidence on the prevailing public health impacts of ETS.

Gray Robertson, president of ACVA Atlantic, Inc., and an indoor environmental scientist, addresses the following basic issues: what is actually in indoor air, and how big a problem is ETS in the context of other indoor pollutants? He finds that ETS is a relatively trivial factor in most indoor environments. In fact, his building measurements indicate that there are indeed potent substances present in indoor environments, such as asbestos, but that ETS is typically not a major substance in indoor environments. Robertson argues that better ventilation systems and not smoking bans are the key way to improve overall indoor air quality. For example, many buildings that have banned smoking have poor ventilation systems based on recycled air. A smoking ban in such a building removes a minor substance in its indoor air while ignoring other major indoor air pollutants. Cleaner indoor air environments require better ventilation systems, not smoking bans.

In chapter 5, the perspective shifts from health and indoor air quality issues to the economic aspects of the ETS issue. Walter E. Williams, a prominent economist at George Mason University, puts the economist's point of view forward by asking who owns the air. In answering this question, he takes a property rights perspective on ETS. Most indoor environments are owned by someone—a factory owner, a restaurant owner, a museum "owned" by the public. In each case there are clear incentives for the owner of the facility to provide the type of indoor environment preferred by patrons. To do otherwise would be tantamount to losing money. Williams thus reasons that government intervention to ban smoking cannot improve upon the environments that already prevail in the private sector. Notice that Williams's argument is not tantamount to a call for laissez-faire for smokers. Some restaurants will ban smoking, and others will not. The point is that government is not necessary to the process of telling people where and when they can or cannot smoke.

The following two chapters are by a labor union official and a president of a public relations firm. In each case, the contours of the argument fit Williams's property rights perspective. René Rondou, secretary-treasurer of the Bakery, Confectionery and Tobacco Workers Union, is concerned about retaining the issue of smoking in the workplace within the framework of the collective bargaining system and about controlling more important indoor air quality problems such as asbestos. Rondou sees public legislation about workplace smoking as unnecessarily giving arbitrary power over workers to management and argues that issues such as workplace smoking are more properly settled in collective bargaining negotiations. Jody Powell, former press secretary to President Carter and currently president of Ogilvy and Mather Public Relations in Washington, D.C., insists that he has sufficient incentive, without government intervention, to provide the indoor environment preferred by his workers. Again, this does not mean that all workplace environments will be smoke free. Some will, and some will not. The point is that the managers of companies will provide the environment that is in the best interests of their workers.

The next two chapters focus on broader social aspects of the ETS issue. R. Emmett Tyrrell, a distinguished journalist and editor of *American Spectator,* argues that the press has sensationalized the scientific case against ETS, aided and abetted by the rhetoric of the surgeon general. He offers some reasons for the press's behavior in this regard and strikes a cautionary note with respect to the treatment of scientific issues by the media. The media want excitement and political conflict, not necessarily sound public policy. Peter Berger, an influential sociologist from Boston University, puts the antismoking movement into larger focus, stressing the reasons that it and its bureaucratic allies in government have seemingly gained the upper hand in the public debate over ETS. Primary among the reasons is the fact that smokers are a poor, politically inactive minority. In a word, it is easier for government to pick on poor people.

Burt Neuborne, professor of Law at New York University, then compares and contrasts the campaign against ETS to prohibition. He finds the argument about third-party costs in the case of smoking unconvincing and argues that the value of personal autonomy ought to prevail in the debate about public smoking. In other words, civil liberties outweigh alleged social costs.

Lord Bruce-Gardyne, member of the House of Lords, provides a perspective on the ETS issue by an inside politician. He stresses the political context, especially in the United Kingdom, of the debate about ETS. He sees it as an old debate between proponents of individual autonomy and those who would restrict individual initiative. He offers some useful strictures for those who lobby politicians in the debate.

Finally, James M. Buchanan, winner of the 1986 Nobel Prize in economics, writes about "Politics and Meddlesome Preferences." His point is simple and profound: we all may find some aspect of others' behavior to which we object. However, if we try to regulate all the things to which some people may object, we end up regulating everything, and we all end up worse off. In other words, a little tolerance goes a long way in a modern society.

I offer some brief concluding remarks in the last chapter.

There is a cogent case to be made, based on a review of the scientific evidence and on the types of arguments made by the authors in this book, that ETS does not pose a huge social issue calling for massive government intervention into the affairs of private citizens. Smokers and nonsmokers will work their differences out, labor unions will bargain collectively with management about smoking on the job, corporations will provide indoor environments that their workers prefer, bars and restaurants will do the same for their customers, and the society and economy in general will go about its business in such a fashion as to mitigate any problems posed by ETS. This does not mean that smokers will have unfettered rights to smoke. To the contrary, their behavior will be constrained by private action. The point is that the heavy hand of government regulation is not necessary in this case. This is, at least, the conclusion that I reach on reading this book.

2

Scientific Issues Regarding Exposure to Environmental Tobacco Smoke and Human Health

Mark J. Reasor

RECENTLY considerable attention has focused on the question of whether exposure to the smoke exhaled by smokers or emitted by their cigarettes (frequently termed *passive* or *involuntary smoking*) is detrimental to the health of nonsmokers. Because nearly 30 percent of the American population are active smokers, the potential for exposure of nonsmokers, at work and at home, to what is termed environmental tobacco smoke (ETS) is considerable. Since a large number of people may be exposed to ETS, the scientific community must make every effort to determine whether such exposure adversely affects health.

In response to this need, many scientists throughout the world are involved in this effort, and much information on this subject has been collected. This chapter presents a perspective on the scientific issues relating to whether exposure to ETS adversely affects the health of nonsmokers. It describes ETS and its similarity or dissimilarity to the smoke inhaled by active smokers (mainstream smoke), examines the nature of exposure to ETS, relates what adverse health effects have been reported, describes problems encountered in analyzing existing studies, and considers what types of studies should help obtain the most scientifically sound and definitive answers. This chapter does not address social or political aspects of this topic.

ETS and Comparisons to Sidestream and Mainstream Smoke

ETS originates principally from the smoke emitted from a cigarette between puffs. The smoke released from the cigarette, termed *sidestream smoke,* consists of particles that are visible and gases that are invisible. Smoke exhaled by the smoker *(exhaled mainstream smoke)* also contributes to ETS, but it is believed to make less of a contribution than sidestream smoke.

As sidestream smoke and exhaled mainstream smoke are released into an indoor environment, a number of processes occur that influence the concentration and composition of the end product, ETS. Overall these events are termed *aging.* The first and most important event from a health perspective is dilution with room air. This will be extensive under most circumstances and will be influenced by a number of factors, including the size of the room and the existing ventilation rate. The concentration of ETS also will depend upon the rate of smoke generation and, hence, the number of smokers in the room. Other important processes affecting the level of ETS include adsorption onto and absorption into materials in the room, such as furniture, drapes, and carpets, as well as chemical and physical reactions that the material undergoes. The types of reactions that occur are not well characterized, but there is some evidence that changes occur that both increase and decrease the presence of potentially toxic substances. In general, the extensive dilution of the smoke materials would be expected to offset or diminish possible health effects resulting from the generation of more toxic species. Although the presence of potentially toxic substances in an environment is cause for concern, exposure to very low levels of a given substance may not result in toxicity even when exposure to a much higher level will.

Because it is much easier to collect freshly generated sidestream smoke in a scientific laboratory than to collect ETS under actual conditions, sidestream smoke has been commonly used as a surrogate for ETS. Although much is known about the chemical composition of undiluted sidestream smoke, serious limitations exist in interpreting the information obtained from such an analysis. When all factors are considered, it becomes apparent that ETS and sidestream cigarette smoke are quantitatively and, to some extent, qualitatively different.

The composition of ETS has been compared frequently to that of the mainstream smoke inhaled by an active smoker. Constituents found in mainstream smoke are also found in sidestream smoke. Under standardized smoking conditions used in the controlled laboratory setting, higher amounts of most of these materials are present in the undiluted sidestream smoke released from a cigarette than are present in the mainstream smoke generated from the same cigarette. From this type of investigation, it has been reported that sidestream smoke has more toxic potential than mainstream smoke. Although this may be true if the undiluted materials are compared, such a statement is misleading since it is ETS and not undiluted sidestream smoke to which nonsmokers are exposed.

Exposure to ETS

A substance may be toxic at one concentration but not at a much lower concentration. Therefore of primary importance in evaluating the potential health effects of any material, including ETS, is knowledge of the dose, or amount, of the material to which the individual is exposed. Because of difficulties in accurately determining exposure levels, interpretation of purported health effects as a result of exposure to ETS is hampered. To illustrate this problem, it is appropriate to compare exposure to and retention of ETS by a nonsmoker to the inhalation and retention of mainstream smoke by an active smoker.

Nonsmokers inhale ETS almost exclusively through the nose with a normal breathing pattern. Smokers draw mainstream smoke directly into the lungs, often with a deep breath and a pause that allow prolonged contact of the material with the lungs. This difference in respiratory pattern increases the chance for deposition and absorption of material by active smokers. Studies have shown that differences in the physical properties of mainstream smoke and sidestream smoke (used in this case as a surrogate for ETS) result in less retention of sidestream smoke particles in the lungs compared to particles of mainstream smoke. Studies on the retention of particles in ETS have not been reported. It is often said that exposure to ETS is the same as actively smoking cigarettes, the only difference being that the components are more dilute. For the reasons already noted, such extrapolation is scientifically inaccurate and therefore unwarranted.

Certain substances in tobacco smoke have been detected in indoor environments. Unfortunately it has not always been possible to determine the contribution, if any, of ETS to indoor air pollution, principally because a number of the components of ETS are not unique to tobacco. For example, particulate matter and carbon monoxide have been used as markers for the presence of ETS. Nontobacco-related particles can arise, however, from both indoor and outdoor sources. Indoor sources include coal- and wood-burning fireplaces and general house dust. Particles from outdoor sources can enter homes or buildings under any circumstances and particularly when the structures are near construction sites, agricultural fields, heavily traveled roads, and so on. In spite of these complications, there is good evidence that the level of particulate matter can be higher in homes when smokers are present. Carbon monoxide is a gaseous material that results from the incomplete combustion of organic material and may arise in an indoor environment from sources other than the burning of tobacco—for example, from gas stoves. This chemical can also arise from outdoor sources. An example is in office buildings where fresh air may be drawn in from sites near the generation of automobile exhaust.

The most specific marker for ETS appears to be nicotine because tobacco is the only source of this chemical. Nicotine has been measured in indoor environments and has been detected in body fluids (blood, urine, and saliva) of nonsmokers. Nevertheless, the use of nicotine as a marker for ETS has certain limitations. Nicotine in ETS is a gas, so it can serve at best only as a marker for components in the gas phase and not for materials in the particulate phase. Consequently it is not possible to determine how much absorption of other substances has occurred nor can information about the cumulative exposure to ETS and accumulation of materials in the body be derived. Proper characterization of the contribution of ETS to indoor air pollution will require an integrated approach that examines a number of components simultaneously.

There is concern about the possible presence of potential carcinogens in ETS. Such material may exist in minute quantities, compared to nicotine, making their detection in body fluids of nonsmokers very difficult. Ideally it would be desirable to have a marker representing a carcinogenic material that could be used to measure the cumulative exposure to ETS and permit an estimation of carcinogenic risk. At present, such a marker is not available, although scientists are searching for one.

Recently investigators have been looking for the presence of mutagenic substances in the urine of nonsmokers as an indicator of exposure to carcinogens that may be present in ETS. (If a substance causes mutations, it is believed to have the potential to cause cancer.) Even if ETS-specific mutagens are found in the urine of nonsmokers, however, the health consequences of their presence would be difficult to interpret.

Health Effects of ETS

Although very little is known about ETS and it is not valid to extrapolate from active smoking to passive smoking because they are markedly different exposure situations, nevertheless there is considerable concern about the effects of ETS exposure on human health. Reports on this topic have been appearing in the scientific literature for well over ten years. Interest has centered on two types of health effects: acute and chronic.

With any substance, acute effects may result from a single exposure and disappear after exposure stops. Chronic effects are those that result when exposure occurs over a longer period of time, generally months or years. Chronic effects may or may not be more severe than acute effects, depending upon the substance and levels of exposure. It is much easier to attribute acute effects to a particular material than it is to attribute chronic effects because chronic effects often appear well after exposure has occurred, making it difficult to identify a causative agent.

Acute Effects

The most common acute responses reported by individuals who are exposed to ETS are detection of unpleasant odors and irritation of eyes, nose, and throat. The perception of an offensive odor may alter a person's sense of well-being and comfort but is generally not considered to be a health effect. Since it can affect attention, attitude, and work performance, however, it warrants concern.

There is little question that some nonsmokers experience irritation of eyes, nose, and throat when exposed to sufficient amounts of ETS, particularly in inadequately ventilated environments. The responses to ETS are variable; some individuals report no irritation, and others claim to be allergic to ETS. It has not been established that such reactions are true allergies, although certain persons do appear to be highly sensitive to the irritative effects of ETS. Although there are reports that individuals with respiratory disorders such as asthma and emphysema have an increased sensitivity to ETS, inadequate scientific evidence exists to confirm this.

Chronic Effects

A number of chronic health effects purportedly resulting from exposure to ETS have been reported in the scientific literature, among them lung cancers and cancers in tissues other than the lung, respiratory disturbances in adults and children, cardiovascular disorders, and certain other adverse effects in children. The studies reporting these effects have been the subject of intense scrutiny in recent years, and there is controversy concerning their conclusions.

In 1986, two major reports on ETS were published, one by the U.S. surgeon general and the other by the National Research Counsel of the National

Academy of Sciences (NAS), both containing an extensive review of the scientific literature by eminent scientists and physicians from throughout the world. They concluded that ETS has not been shown scientifically to increase the risk of cancers in tissues other than the lung or to increase cardiovascular disorders or deficits in respiratory function in adults, although they claimed a positive association between exposure of the nonsmoker to ETS and an increased risk of lung cancer. Additionally, both concluded that young children whose parents smoke have an increased frequency of certain conditions, including ear infections, acute respiratory illness, and respiratory infections in early infancy and an increase in cough, wheezing, and sputum production. Other commentaries and reports have addressed these issues of the risk of lung cancer in nonsmokers and health effects in children, some agreeing with the finding of a positive association and others disagreeing. It is important to examine why this controversy exists.

Epidemiological investigations have been used to examine the relationship between ETS exposure and lung cancer in nonsmokers. It has been reported that nonsmoking wives of men who smoke have an increased risk of about thirty percent in developing lung cancer compared to nonsmoking wives of men who do not smoke. Because of the nature of epidemiological studies, it is very difficult to establish a direct association between exposure and disease and is virtually impossible to establish causality. Critical evaluation clearly shows that a number of deficiencies exist in studies reporting an increased incidence of lung cancer in nonsmokers as well as those reporting no increase.

The most serious problem is the absence of verified exposure information. In no study to date is there an accurate or adequate measurement of exposure of any of the subjects to ETS. Exposure is most frequently estimated from the smoking habits of the spouse, and this information is obtained by questionnaire or interview with the subject. Such methods have serious limitations, not the least of which is the accuracy of respondent recall of facts. Evidence exists that spousal smoking habits may not accurately represent ETS exposure. Additionally the studies suffer from one or more of the following problems: an inadequate number of subjects to permit statistical analysis, inaccurate or unverified information about the subjects, lack of clinical verification of the actual presence of lung cancer, and misclassification of the subjects with regard to their smoking status or exposure conditions. These flaws need to be corrected for in any future epidemiological study before reliable conclusions can be drawn.

There are also substantial doubts about the appropriateness of aggregating or combining flawed epidemiologic studies in an effort to reach conclusions that cannot be supported or confirmed by the studies individually. The bringing together of individual studies does not necessarily diminish the importance of the underlying study defects, and the aggregation technique thus may result in spurious conclusions. The overriding need in the area of ETS and lung cancer is the design of better and more rigorously conducted investigations, not the statistical manipulation of studies that already have been conducted.

Since sidestream smoke contains carcinogens, it is likely that ETS does also; however, the concentrations of these materials would be expected to be very low in ambient ETS and, if present in the particulate phase, would be even less likely to be retained in the lungs. The important issue is whether nonsmokers are exposed to a sufficient amount of such materials to cause lung cancer. Possibly this question may never be answered definitively. What is clear, however, is that the investigations completed to date have not offered satisfactory answers. It is imperative that research continue on this topic.

Because young children are less able than adults to control their exposure to ETS and because exposure in youth may impair normal development, the issue of whether exposure to ETS results in health problems in this age group is significant. Nicotine and its metabolite have been detected in the body fluids of children of smokers. Since no quantitative cumulative measurements of ETS exposure have been obtained in the studies where health effects have been reported, the extent of exposure cannot be verified. Nevertheless, there are a number of reports that a higher incidence of the disorders described occurs in young children of parents who smoke. Some scientists have questioned, however, whether this is actually due to ETS, since other factors may contribute to these responses. For example, cross-infection from a parent may have been the causative agent rather than ETS. In addition, socioeconomic factors and the possible effects of smoking during pregnancy have not always been considered in these studies. Because of the sensitive nature of this issue and the possible important consequences of the findings, researchers need to design appropriate studies to eliminate these deficiencies.

Future Needs of Scientific Studies

There is no consensus among scientists on the question of whether exposure to ETS can adversely affect the health of nonsmokers. Further research is needed to determine whether ETS presents health risks.

The fundamental need is to develop quantitative methods to measure the cumulative exposure of the nonsmoker to ETS. Unfortunately it is not clear how this will be done in the near future. Organizations are developing comprehensive questionnaires designed to overcome many of the shortcomings of questionnaires used in previous studies; nevertheless, interpretation of studies using questionnaires of any kind will continue to be limited.

Experiments where human subjects are exposed to ETS in a laboratory setting permit better control over exposure conditions; however, such studies can be conducted only on a short-term basis and therefore will not answer questions about the more important issue of chronic exposure and the possible consequences to health.

The present data do not show a clear association between ETS and lung cancer, much less meet the criteria for judging causality. Because of the small increases reported in the risk of lung cancer, it is not likely that causality can ever be established using epidemiological techniques. Consequently new experimental approaches are needed. One approach that warrants attention is the use of laboratory animals for short- and long-term exposure to ETS. Most information on the toxicity of nonmedicinal chemicals has been obtained in animal studies, so it is logical to utilize such an approach in the study of ETS. It should be possible to expose laboratory animals to ETS for periods up to their entire lifetime. Under controlled circumstances, the development of lung cancer in adult animals and the development of respiratory disorders in newborn animals, as well as many other toxicities, can be assessed. Such studies have obvious difficulties, not the least of which is their expense. Additionally although it is important to determine if there is a potential for toxicity, as lifetime exposure to high concentrations of ETS would permit, it is necessary to conduct experiments that simulate realistic conditions of ETS levels and durations of exposure.

As the quest continues to evaluate whether ETS is a health hazard, scientists must strive to conduct carefully designed and appropriate research. Careful scientific evaluation and dispassionate criticism of the studies must occur. Published data must be reassessed by other scientists to evaluate the validity of the conclusions. Results or conclusions should not be presented or interpreted with a preconceived bias on the part of the investigator or reader. Through such a process, confidence in reported findings, either of a positive or negative association between ETS and human health, can be established, and appropriate information can be presented to a concerned public.

3

Complexities in Developing Public Health Programs

A Public Health Consultant's View

W. Allan Crawford

What Is Public Health?

Public health measures are, in theory, an expression of the will of the people and consequently purportedly for the ultimate benefit of all society. Generally such a measure is the subject of legal statutes by governments, which of necessity lead to restrictions on the activities of individuals, communities, industries, and life-styles. Requirements are also imposed on society to provide money and methods to achieve the targets that a population has accepted as being desirable, beneficial to all, practicable, and politically feasible.

Members of the health professions, or other professions, or simply observant citizens can only advocate new measures or the removal of old ones. Public health authorities have no independent basic power; they have only the power to apply the laws related to health.

To many people, public health relates only to such essential needs as water supplies, sanitation, control of infectious diseases, housing standards, food supplies, and provision of health care facilities. This was true of the eighteenth and nineteenth centuries. Although these activities have been maintained in this century, there has also been a major evolution in other areas affecting human well-being. Now offspring of public health departments are to be found in the science and practice of occupational health, air, land, and water pollution control, community health, and medical and scientific research units supported by government (public) funds.

The director of a public health service is a servant of the public and will be presented with proposals on how to meet the World Health Organization's definition of health as a "state of complete physical, mental and social well-being." The director's task is to advise government on such proposals and in so doing must take into account many factors. This chapter considers the questions raised by the issue of environmental tobacco smoke (ETS) and health as an example of an area where these public health considerations must be taken into account. This is a subject of considerable debate, argument, and emotion, and it represents a divergence of medical and scientific opinion.

Basic Questions

The first question to consider is that concerning the plausibility of the hypothesis that ETS may have an effect on nonsmoker health and well-being. The second question relates to practicality. Can the restriction or prohibition of ETS be achieved, and, if so, how confident can one be that a reduction of disease, exposure, or a positive health benefit would result? The third question relates to the effect on the body politic as a whole and on groups of individuals within the whole, always bearing in mind that public health measures have sociopolitical aspects and that individuals within the groups vote and that this will be a consideration of ministers or secretaries of state for health and their political colleagues.

Moreover, a professional trained and experienced in public health has to take into account not only the hypotheses, the clinical reports, and the epidemiological data on an issue but also the nature of disease and the possibility that the diseases of relevance might have risk factors associated with them. Taking this necessity into account, consideration of these three basic questions reveals the following:

Question 1: It is possible that ETS causes disease in nonsmokers because it does contain substances associated with disease. Detailed quantification, however, is lacking and dose-effect data are not conclusive.

Question 2: Although presenting major difficulties, restrictions or prohibitions are practicable; however, data are inadequate—practically nonexistent in fact—on the health of the general public exposed to ETS. In addition, the potential effects of ETS are confounded by indoor and outdoor environmental pollutants.

Question 3: Two large segments of the public are involved, smokers and nonsmokers.

Such answers lead to the conclusion that this complex issue is important and worthy of study. The next move from a public health viewpoint is to examine the basic publications on reported health effects, the reviews and critiques of these papers, and authoritative publications and consensus opinions of experts.

Reported Health Effects

The health effects associated with ETS exposures include irritation of the eyes, nose, and throat, respiratory symptoms in children and adults, exacerbations of allergy and asthma and coronary artery disease, as well as lung and other cancers. The U.S. surgeon general, the U.S. National Research Council (NRC) Committee, and the World Health Organization International Agency for Research on Cancer (IARC) have reviewed the available data. Their reports, which run 300 to 400 pages and provide in excess of 1,000 references, indicate that the volume of papers and numbers of publications on ETS is large and the results conflicting.

Specificity and Confounding Factors

Of prime importance in assessing the need for public health action is the specificity of the relationship of ETS to the complaint or disease and the presence of other related factors. These include not only such substances as viruses, bacteria, fungi, chemicals, and physical agents but also the impact of inexact and inaccurate responses to questionnaires used by epidemiologists in their surveys. The extent of these factors is illustrated in table 3–1, from the NRC report.

Accounting for these factors is a formidable task for any research group. In addition, there is the onerous task of acquiring a sufficient number of subjects and controls. Without adequate case numbers, determining the relevance of the research findings to a wider population is difficult. Even when efforts are made to increase the numbers by combining data from separate studies, the results are at best speculative, if not invalid.

Table 3–1
POTENTIALLY CONFOUNDING AND EFFECT-MODIFYING FACTORS IN EPIDEMIOLOGIC STUDIES OF EXPOSURE TO ENVIRONMENTAL TOBACCO SMOKE

Unreported active smoking
Tobacco products
Marijuana
Clove cigarettes
Developmental factors
Maternal smoking during pregnancy
Factors related to outdoor environment
Outdoor temperature, humidity
Respirable and nonrespirable particles, such as, fugitive dust
Pollens and other allergens
Factors related to indoor environment
Crowding
Number and age of siblings
Total number of people and animals in dwelling unit
Total number of smokers in dwelling unit
Household conditions
Frequency of air exchanges
Temperature and humidity
Use and condition of air-conditioning unit
Conditions of child care facilities
Unvented combustion products from heating and cooking stoves
Respirable and nonrespirable particulates, such as wood smokes
Pollens, molds, mites
Allergens and infectious organisms
Formaldehyde
Factors related to work and hobbies
Work- and hobby-related exposure to gases, fumes, particulates
Miscellaneous factors
Annoyance response to tobacco smoking
Reporting biases

To justify a public health measure that affects in one way or another the majority of people at home, work, recreation, and general daily activities, the data available as the basis for governmental consideration must be reasonable, if not compelling.

The literature on health effects from exposures to ETS on children 1 year old or under is suggestive, but such studies lack the sophistication to assess properly a true effect or its magnitude. The data on respiratory health and lung function for children and adults in general are not adequate, and experimental studies on asthmatics are too few and the results conflicting. All of the studies have serious shortcomings that are noted in the major reviews. Geographic, climatic, and urban-rural factors have to be considered, as do socioeconomic aspects.

With regard to lung cancer, the data are generally restricted to studies of spouses of smokers and for the most part do not reach statistical significance. There is a division of views on whether the small nonsignificant increase in risk reported in some studies could be explained by the misclassification of as few as 5 percent of the subjects as nonsmokers when in fact they are or were smokers. Furthermore, in many of the studies of cancer in spouses of smokers, autopsy data are not available; thus not only is it possible that the cancer identified as originating in the lung was secondary to cancer of another site but that cancer may not even have been present.

Furthermore lung cancer in nonsmokers in the general public is a rare disease. Increased relative risks of 10 to 50 percent for exposed nonsmokers have been calculated. This would not be discernible even in large studies and could, if found, be attributable to exposure to many known (and not known) environmental pollutants in the workplace and other locations.

The limited data on ETS and coronary artery disease in the form of angina are at best questionable and need further study.

With regard to other health aspects, such as pregnancy and children, growth rates, cancer of organs other than the lung, and effects on basically defective lungs, research would be required to explore any proposed hypotheses or claims of disease causation.

In summary, studies of an epidemiological nature do not provide compelling data on exposure to ETS and human disease. Thus the types of data discussed in this review in relation to justifying possible public health legislation are far from adequate for that purpose.

Confounding Environmental Factors of Major Importance

Outdoor Air Pollution

Public health authorities have strongly relied on the work and research of their colleagues in pollution control and occupational health systems in reaching conclusions about outdoor air pollution. They have been helpful in collecting highly suggestive data from industrial Britain. In addition to the adverse aesthetic effects of air pollution, there is compelling evidence regarding its impact on the lungs of urban residents. This was dramatically displayed in industrial Britain, the Meuse Valley in France, the Donora Valley in the United States, and Tokyo in Japan. The better-publicized events illustrating the potential impact of air pollution occurred in the 1950s. During that period, thick fogs were common and were shortly followed by outbreaks of bronchitis and increases in mortality from cardiovascular and respiratory disease. The acute effects reported by the public precipitated the political actions that have been successful and continue to be successful through the application of clean air acts.

Although actions to limit the acute effects were successful, the public gave little attention to the possible long-term effects in the forms of chronic obstructive lung disease (COLD) and lung cancer. Few noted that COLD was largely an urban disease, that cancer mortality in cities was twice that in rural areas, and that there were other geographic variations in disease rates. Fewer still noted the evidence from the post–1939–1940 era that disease patterns changed among migrant populations. For example, the incidence of British lung diminished among those moving from Britain to less polluted places, such as South Africa, Australia, or New Zealand, and diminished even more among the offspring of those migrants. The national and international data certainly suggest a relationship between air pollution and lung disease and cardiopulmonary disease.

Indoor Air Pollution

For more than a hundred years, public health authorities have advocated adequate ventilation in the home and workplace to dilute the pollutants, notably bacteria, that cause disease; the results suggest that such measures were effective. In the last two decades, however, the energy crisis led to a diminution in workplace and home ventilation. The adverse effects are described in chapter 4. In many countries, conflicting advice was (and continues to be) given by the relevant officials in that energy authorities suggested insulating buildings to diminish thermal needs while health authorities maintained the need for good ventilation. Despite their support for adequate ventilation, most public health authorities were unaware of the concentrations of hazardous chemicals emitted by heating and cooking appliances, and once again the initiatives of pollution control scientists revealed an overlooked problem of basic importance.

People today spend 80 to 90 percent of their life indoors. Manufactured materials, floor covers, furniture, particle boards, plastics, glues, stoves, and so on contribute to indoor pollution, as do open fires and cooking, and some of these pollutants are potential human carcinogens. The question arises: Do these chemicals contribute to the incidence of cancer?

In the 1960s and 1970s, these indoor pollutants were mainly considered to be of academic interest, but when environmental control agencies in Sweden and the United States found that a known human carcinogen of natural origin was being trapped in inadequately ventilated buildings, action took place. The human carcinogen was radon, a gas in the radioactive decay system from naturally occurring uranium, and it was seeping into buildings and homes. Radon decay products, once known as radon daughters, emit radioactive alpha particles, which are carcinogenic to the lung. The U.S. Environmental Protection Agency has issued official warnings and advice. Their official estimate is that this radiation may induce 5,000 to 20,000 lung cancers per year in the United States.

The discussion clearly illustrates the dilemma facing health authorities with regard to advising governments on the subject of ETS and human health. How is it possible to take into account the many confounding factors in the assessment of disease causation and policy development?

ETS in Perspective

Public concern about air pollution, although certainly understandable, is frequently in inverse proportion to the concentration of the pollutants present in the environment. In some quarters, there are concerns about the presence of minute amounts of various substances in the air. At times such concerns have seemed almost irrational, failing to consider that the improvements in health and longevity have occurred in the presence of these low levels of pollution. If clean air is to be defined as air free of all pollutants, all industry, all communal activity, and, indeed, breathing itself will have to be halted.

The current lack of perspective with regard to ETS and other pollutants is deplorable because it could lead, by political pressure, to the adoption of draconian measures against a large section of the public, despite the absence of incriminating evidence. In this regard, Patrick Lawther of the United Kingdom writes: "The lack of sense of history and perspective often accompanies the most laudable zeal, but the distortion which follows careless, albeit passionate, thought and clamor for action will impede progress in the search for the real effects of pollution and the implementation of measures to abate it. We must be on our guard."

Ministers of health and their equivalents in government should be informed that nonsmokers are exposed to small amounts of ETS under typical conditions without apparent health effects. Some of the substances present in ETS have been classified as carcinogenic but not in the quantities found in ETS. With the exception of nicotine, those same substances occur in the general environment and in the home as a result of burning fossil fuels and other day-to-day activities. Any increased risk of lung cancer to non-smokers exposed to ETS would appear to be low, based on the epidemiologic evidence, and generally not statistically significant. Nor can the influence of confounding factors be excluded. Although it is common practice to infer that there is no evidence of a threshold for carcinogenesis, most toxicologists are of the view that it would be a waste of resources to seek such a threshold or even impossible to establish one.

Clearly in the area of ETS, knowledge is limited, and further research is needed. This must be kept in mind in considering public health measures that could affect the life-styles of millions of people. This is obviously not the time to introduce legislation. Rather it is the time to inform the public in a dispassionate manner of what is known and what is not known and to review and take account of new data as they are produced.

4

Building-Related Illnesses

Tobacco Smoke in Context

Gray Robertson

Historical Background

Concern for indoor air quality has a relatively long history, but control for indoor air quality is still in its infancy. The Romans, long before the coming of Christ, were careful of how they disposed of their rubbish to avoid undue pollution in their living areas. More recently, Benjamin Franklin expressed his concern and anticipated ours when he wrote to a friend who was a physician to the emperor of Vienna: "I considered fresh air an enemy, and enclosed with extreme care every crevice in the rooms I inhabited. Experience has convinced me of my error. I am persuaded that no common air from without is so unwholesome as the air within a closed room that has been often breathed and not changed."

Franklin had come to realize that air inside a room needs to be changed. Essentially any room that starts off with fresh air rapidly undergoes change when that room is inhabited by people. People's bodies raise the room temperature, and their perspiration increases the moisture content. Skin scales are shed, and body oils and vapors migrate into the room air. As we breathe, oxygen is taken in, and carbon dioxide gas is exhaled. The air becomes stale. The rate of this adverse change in air quality is proportional to the number of people in the room and the level of their activities. This stale air must be removed and replaced with fresh air by ventilation. The prime goal of ventilation is to reduce contamination by dilution, a fact sadly overlooked in modern society. However, there are mitigating circumstances; dilution of the indoor air can work only if the outside fresh air is clean.

For many years, this outside air was the major focal point of pollution. As recently as 1950 through the 1960s, the rapid growth of population and industry and the increasing use of automobiles led to the fogs and photochemical smogs that invaded major cities. Since such excessively contaminated outdoor air was being brought indoors, it is hardly surprising that the use of air-conditioning found such a receptive market.

The basic concept of air-conditioning is to provide filtered, clean air at the most comfortable temperature range to the occupants of a building. In time, most of the major buildings throughout the United States were fitted with air-conditioning. Everyone took it for granted.

Then, in July 1968, an explosive epidemic of illness characterized principally by fever, headaches, and muscular pains affected at least 144 persons, including 95 of 100 persons employed in a health department building in Pontiac, Michigan. A defective air-conditioning system was implicated as the source and mechanism of spread of the causative factor, but extensive laboratory and environmental investigations failed to identify the etiological agent so the incident was simply labeled Pontiac fever. Many years later, it was learned that this illness was caused by a bacterium now identified as Legionella pneumophila. This name was first attributed to the causative organism of the 1976 outbreak of a disease resembling pneumonia that struck 182 people attending an American Legion Convention in an air-conditioned hotel in Philadelphia; 29 of those affected died. The Center for Disease Control in Atlanta has estimated that this same bacterium strikes between 25,000 and 45,000 persons each year in the United States alone.

In retrospect, we can now identify the tragic events of the first Legionnaire's epidemic as a starting point of a new focus on air contamination, namely indoor air pollution.

Let us review the evolution of recent ventilation standards and consider their impact on indoor pollution. Until 1973, the ASHRAE (American Society of Heating, Refrigeration, and Air-Conditioning Engineers) standard for the correct ventilation of offices was a minimum rate of 15 cubic feet of fresh air per minute (cfm) per person, with a recommended rate of 15 to 25 cfm.

In the early 1970s, the costs of oil started their dramatic increase, and a new science of energy conservation was born. Ventilation rates were reduced, and lower volumes of fresh air were introduced into the buildings.

A somewhat arbitrary figure of 5 cfm was thought to be satisfactory. However, people exposed to only 5 cfm of fresh air noticed a new, widespread phenomenon, tobacco smoke-filled rooms. Thus, from the early 1970s, tobacco smoke became synonymous with indoor pollution.

Obviously it was undesirable to work in a smoke-filled room, and the solution led to increased ventilation. Unfortunately we reacted only to the visible pollutant, the tobacco smoke. Ventilating engineers estimated that it would require up to 25 cfm of fresh air to remove heavy accumulations of smoke, and in 1981 ASHRAE published its dual ventilating standard (ASHRAE 62-1981) calling for 5 cfm of fresh air per person in a smoke-free environment and 25 cfm if smoking was present.

This standard resulted in several innovations. Some companies, now obsessed with energy savings, introduced smoking bans in their buildings. Architects and building designers were confronted with a difficult problem: how could a building served by only one ventilating system cater to both smokers and nonsmokers? Some took averages. If only one-third of the population smoked (⅓ at 25 cfm) and thus two-thirds did not (⅔ at 5 cfm), the resulting rate would be 11.67 cfm.

Meanwhile, other events occurred with increasing regularity. People working in offices with only 5 cfm, even in buildings where smoking was never permitted (such as some hospitals, schools, and commercial offices), started to get sick. The so-called sick building syndrome appeared. The biggest single common denominator in most of these "sick buildings" was that carbon dioxide levels were high. Readings in excess of 1,000 parts per million (ppm) of carbon dioxide indoors were compared with average outdoor or fresh air values of 350 to 400 ppm. In short, inadequate ventilation was the main cause. If the ventilation is inadequate, all forms of indoor pollution can accumulate in the air inside the building.

These facts sent the code committees back to the drawing boards, and as recently as December 1986, ASHRAE issued a proposed new standard for ventilation in offices of 20 cfm per person. No distinction is made for smoking or nonsmoking, and we have gone full circle back to the era when the sick building syndrome was unheard of and when complaints of passive smoking were never a focus of attention (strange, since smoking has been around since Columbus brought tobacco seeds from the New World in 1492).

In fact, the visibility of smoke has long been utilized by ventilation specialists to evaluate the efficiency of an air distribution system. Smoke tubes are released and air current identified by the movement of smoke.

The rate of smoke dissipation was proportional to the efficiency of ventilation. If the smoke was trapped, the ventilation was inadequate, and whenever this occurred, all indoor pollutants, visible and invisible, were similarly trapped.

Indoor Pollutants: The Sources

Virtually everything we use in the interior sheds some particulates and/or gases. When a building is new, some compounds are given off quickly and soon disappear. Others continue "off-gassing" at a slow pace for years. Common office supplies and equipment, especially duplicators and copiers, have been found to release dangerous chemicals, and we have even found formaldehyde being released from bulk paper stores.

People themselves are a major contributor. Each person sheds literally millions of particles, primarily skin scales, per minute. Many of these scales carry microbes, but fortunately the vast bulk of these microbes are short-lived and harmless.

Clothing, furnishings, draperies, carpets, and so on contribute fibers and other fragments. Cleaning processes, sweeping, vacuuming, and dusting normally remove the larger particles but often increase the airborne concentrations of the smaller particles.

Cooking, broiling, grilling, gas and oil burning, smoking, and coal and wood fires also generate vast numbers of airborne particulates, vapors, and gases. If the windows and doors are closed, all of these accumulate in that internal environment.

Classification of Indoor Pollutants

Perhaps the simplest classification is the division of pollutants into gases and vapors, both organic and inorganic; fibers and dusts, which can be subdivided into total suspended particulates (TSP) and respirable suspended particulates (RSP) (the latter being the more important since these are the particulates that can pass through the natural filters of the nose and enter the lungs); and microbiological organisms, which can be viable or nonviable fragments.

Gases and Vapors

Organic Chemicals. These are arguably the widest range of pollutants, with literally thousands of specific types, fortunately occurring in very dilute concentrations, usually expressed as parts per million or per billion. Most of these are presumed to be safe at the very low levels encountered, although some synergism between different organics or some incidences of organics sensitizing people to other pollutants cannot be ruled out. Usually the organics are more a problem in the typical home than in the office, and concentrations in the home are usually higher than in the office, mainly due to lower air exchange rates.

Two organics are of particular note: formaldehyde and methylene chloride. Formaldehyde has received considerable attention due to its widespread presence in such substances as adhesives, glues, and urea-formaldehyde insulation. Most of the documented case histories of severe formaldehyde pollution are in homes, especially insulated mobile homes, where the concentration of formaldehyde-containing materials per unit area is higher than in typical offices, and ventilation rates are lower.

A potentially more serious pollutant is methylene chloride. This compound has been shown to be carcinogenic in rats and mice when inhaled. With sublime ignorance, many of us have sprayed a concentrated form of this material over our heads each morning as hair spray, and its widespread use in spray paints and insecticides is neatly disguised on the product labels as chlorinated solvents or aromatic hydrocarbons.

Radon Gas. Radon, a decay product of uranium, is present in variable quantities in soils. It moves from the soil by diffusion into the soil's air pockets or into soil water. Then the radon can migrate from the soil air through unvented crawl spaces, building foundation cracks, and so on into the indoor space. Some building aggregates, cinder block, and other materials also contain radon, and out-gassing from these materials adds to the indoor air levels. In other cases, radon enters a building through the water supply. Some of this radon is released when there is turbulence of the water, such as a running tap. It has been estimated by some researchers that anywhere from 10 to 15 percent of the average radon we are exposed to comes from such water. The general consensus, however, is that the principal source of radon in buildings undoubtedly is the soil gas. Pollution by radon is far more prevalent in homes than in offices, again mainly due to the lower air exchange rates in homes plus the fact that homes have a larger area of exposure to soil relative to building volume and soil leakage area.

Inorganic Oxides. Carbon dioxide is produced by respiration and combustion. Oxides of nitrogen and sulfur are combustion products associated with gas stoves, wood and coal fires, and kerosene heaters. Carbon monoxide is emitted from unvented kerosene heaters or wood stoves, and it frequently diffuses into buildings from automobile exhaust fumes generated in adjacent garages. Only very small to trace quantities of each of these gases and other organics are present in cigarette smoke.

Ozone is another gas that is generated, usually in very small quantities, by miscellaneous copying machines and by certain electrostatic precipitators used to clean the air. In one specific case we studied, the maintenance staff of a building switched off the main air supply fans over the weekend but omitted to switch off the central electrostatic precipitators. Thus ozone accumulated inside the air handlers and was subsequently delivered each Monday morning. When the fans were switched on, the people working in the areas involved experienced severe, but temporary, discomfort.

Fibers

Asbestos. Prior to 1973, asbestos was the material of choice for fireproofing, thermal insulation, and sound insulation. It was used as a spray-on insulation of ceilings and steel girders; as a thermal insulation of boilers, pipes, ducts, and air-conditioning units; as an abrasion-resistant filler in floor tiles, vinyl sheet floor coverings, roofing, and siding shingles; as a flexible, though resistant, joining compound and filler of textured paints and gaskets; as a bulking material with the best wear characteristics for automobile brake shoes; and in countless domestic appliances such as toasters, broilers, dishwashers, refrigerators, ovens, clothes dryers, electric blankets, and hair dryers.

Many asbestos-bearing materials or products are of no health risk whatsoever when used in the normal course of events. However, if for any reason of wear, abrasion, friability, or water damage, any of the asbestos fibers are released into the air and inhaled into people's lungs, there is a health hazard. The scientific evaluation of all available human data provides no evidence for a safe level of airborne asbestos exposure; thus any quantity should be considered potentially dangerous.

Glass Fibers. The glass fiber (usually referred to as fiberglass) industry is in its infancy compared with asbestos, and since asbestos-related illnesses manifest themselves only tens of years after exposure, there are some schools of thought that suggest glass fiber fragments will also accumulate in the lungs and cause later problems. This may be so, but the problems are unlikely to be anywhere near as severe as with asbestos. The fibers of glass are not shed in such large quantities as asbestos, and most of the resins bonding the fibers together appear to be extremely effective and long lasting. However, some fragmentation does occur, and this is especially noticeable when the loose fiberglass insulation, popularly used in attics and ceiling voids, is disturbed. Most people have experienced itching on contact with fibrous glass, and dermatitis-type reactions are not infrequent due to airborne fibrous glass particles.

Microbes

Reviews of the literature have shown that the one area of indoor pollution that has received least study or research has been contamination due to microbes. Table 4–1 shows the dominant types of bacteria that my company, ACVA, isolated from inside air-conditioning systems. Eleven (9 percent) of the first 125 major buildings we studied exhibited high levels of potentially pathogenic or allergy-causing bacteria, including Actinomyces and Flavobacterium species. In addition, Legionella pneuophila, the cause of the Legionnaire's disease, has frequently been isolated from inside air-conditioning systems.

Perhaps more significant, we found over twenty-seven different species of fungus contaminating air-handling systems (see table 4–2). Of the 125 buildings we studied between 1981 and 1985, 39 (31 percent) were found to contain high levels of potentially pathogenic or allergy-causing fungi (table 4–3), including Alternaria, Aspergillus, Cladosporium, Fusarium, and Penicillium species. In several buildings with excessive staff complaints, Aspergillus and/or Cladosporium species of fungus were found. In some investigations, epidemiological tests run by various doctors have confirmed severe allergic reactions to the spores of these fungi in all affected staff.

Table 4–1
BACTERIA ISOLATED FROM AIR-CONDITIONING SYSTEMS

Acinetobacter sp.	Actinomyces sp.	Bacillus sp.
Coliform sp.	Chlamydia sp.	Diphtheroids
Flavobacterium sp.	Klebsiella sp.	Legionella sp.
Micrococcus sp.	Nocardia sp.	Nonfermenting gm-ve rods
Proteus sp.	Pseudomonas sp.	Serratia sp.
Staphylococcus sp.	Streptococcus sp.	

Table 4–2
FUNGI ISOLATED FROM
AIR-CONDITIONING SYSTEMS

Alternaria sp.	Aspergillus sp.	Aureobasidium sp.
Candida sp.	Cephalosporium sp.	Chaetomium sp.
Chrysosporium sp.	Cladosporium sp.	Curvularia sp.
Diplosporium sp.	Fusarium sp.	Helminthosporium sp.
Monilia sitophila	Monosporium sp.	Mucor sp.
Mycelia sterila	Oospora sp.	Paecilomyces sp.
Penicillium sp.	Rhoma sp.	Rhizopus sp.
Rhodotorula sp.	Saccharomyces sp.	Scopulariopsis sp.
Streptomyces sp.	Tricothecium sp.	Verticillium sp.
Yeasts		

Subsequent cleaning and removal of the sources of these fungal contaminants resulted in a dramatic reduction of staff complaints.

Dirt in Ductwork

Heating, ventilation, and air-conditioning systems also have been found to be poorly designed and negligently maintained. Excessive dirt accumulations are common in ductwork, even in hospitals. Frequently dirt is built into the systems during construction since the ducts are installed long before the windows are, and construction dusts from the site, plus wood shavings, lunch packets, cans, and other debris, find themselves brushed into the vents. Thereafter over the life of the building, more dirt enters with the supply and return air. Good filters reduce the rate of this accumulation, but the only perfect filter would be a brick wall. All filters, even the ultraefficient ones used in hospital operating rooms, allow fine particles through. Many of these fine particles coalesce, sticking to each other by adhesion or electrostatic attraction, and larger particles grow with time. In commercial buildings, much cheaper and far less efficient filters are common. Many will stop birds and moths, but that is about all. Occasionally filters are omitted, and frequently they are undersized, resulting in large air gaps that allow massive volumes of air bypass to occur. Large electrostatic precipitators theoretically provide ultraefficient air, but in one major building, we found sixteen of eighteen precipitators were inoperative due to broken parts; many had not worked for over a year. In a major hospital, we found the power pack was missing from one of these units. Inoperative electrostatic precipitators provide no filtration.

Table 4–3
ACVA EXPERIENCE, 1981–1986
(Excluding Asbestos)

Total number of major building studies	175
Total number of square footage	36 million
Estimated number of building occupants	200,000
Major problems found	
Inadequate ventilation	64%
Grossly inadequate filtration	30%
Excessive contamination of ventilation systems	38%
Poor relative humidity	11%
Significant pollutants encountered in indoor air	
Allergenic fungi	30%
Allergenic/pathogenic bacteria	9%
Fibrous glass	6%
Tobacco smoke	4%
Motor vehicle exhaust fumes (carbon monoxide)	4%
Organic chemical vapors	3%
Ozone	1%

Dirty ductwork is a perfect breeding ground for germs. It provides an enclosed space, constant temperature, humidity, and food, which is the dirt. No germ could wish for more!

The extent of this potential problem is huge, and it is very surprising what we have found in ducts. Dead insects, molds, fungi, dead birds, and rodents are common. In 1984 we found two dead snakes in air supply ducts. We have found rotting food, builders' rubble, rags, and newspapers. All of these contaminate air. It is the dirt that encourages germs to breed, germs that cause infections.

The dirt and dusts also may be allergenic. In fact, most of the dusts are, by definition, household dusts, which are notorious for causing allergies in many people.

In a recent survey of a 750,000-square-foot hospital in Virginia, we found fourteen miles of ductwork. In that maze of ducts we found: smoke detectors blocked by dirt and inoperative; fire dampers jammed open by dirt; reheat coils blocked by dirt sealing off the fresh air supply; turning vanes and even exhaust grilles sealed with dirt accumulations (in an operating suite, the exhaust fan was still working against these duct blockages, causing such immense negative pressure in the ducts that the ducts were bowing inward almost to the point of collapse); and huge excesses of bacteria and fungi present inside the air handling chambers and throughout the ductwork. Cross-infection rates were high in this hospital, and nurses, doctors, and patients complained about poor air quality. We have since cleaned all the air handlers and the fourteen miles of ducts and have overseen the installation of more efficient filter systems. The hospital ventilation system has been dramatically improved, and its air quality is now well above average.

Symptomology of Indoor Air Pollutants

In general, when one hears of a polluted building or a so-called sick building, one hears familiar symptoms from occupants: eye and nose irritation, fatigue, coughing, rhinitis, nausea, headaches, sore throats, and general respiratory problems. Without doubt, the pollutant most often blamed for these symptoms by the public is environmental tobacco smoke (ETS). However, there are usually confounding variables presented by a number of potential contaminants that preclude a quick analysis establishing a single source of contamination. The main problem is the incredible similarity of symptoms from widely different irritants or even environmental conditions. For example, identical symptoms have been reported for individuals exposed to formaldehyde, ammonia, oxides of nitrogen, and ozone. In addition, similar symptoms are reported by individuals suffering allergic-type reactions to numerous dusts and to microbial spores such as Aspergillus, Penicillium, and Cladosporium fungi, among others. Similar symptoms have been reported from exposure to cotton dust and fibrous glass fragments, and an ever increasing and similar problem is encountered due to low relative humidities. The latter is well known to frequent flyers of airliners where relative humidity levels are frequently as low as 5 to 10 percent, compared to a normal lower comfort level of, say, 40 percent.

This similarity of symptoms is usually unappreciated by the public, and in part it accounts for a bias against tobacco smoke, which is the sole visible air pollutant. We refuse to rely upon or otherwise use the information generated by subjective building occupant questionnaires because of their unreliability. Only upon careful investigation of the entire indoor environment and ventilation system of a building can informed conclusions be drawn about the various causes of poor indoor air quality. As a result, we have made it our business to perform such investigations. Although tobacco smoke is the main suspect of the occupants in many of the buildings we have examined, we have determined high levels of environmental tobacco smoke to be the immediate cause of indoor air problems in only 4 percent of the 175 major buildings we investigated between

1981 and 1986 (table 4–3). This result has been corroborated. In a similar study of 203 buildings from 1978 to 1983, the National Institute for Occupational Safety and Health (NIOSH) found that only four of the buildings studied (2 percent) had indoor air quality problems attributable to high concentrations of tobacco smoke (table 4–4). Significantly, in those few cases where high accumulations of tobacco smoke have been found, ACVA also has discovered an excess of fungi and bacteria in the ventilating system. These microorganisms usually are found to be the primary causes of the complaints and acute adverse health effects reported by building occupants.

Ventilation and Indoor Pollution

The accumulation of many pollutants is a symptom of a more serious problem: inadequate ventilation. In the analysis of the NIOSH studies (table 4–4), approximately 50 percent of the sick buildings were found to be inadequately ventilated. Improper ventilation can sometimes be carried to extremes. Throughout the last four years, before our analysis, the fresh air dampers were closed completely in over 35 percent of those buildings studied by ACVA. Three years ago, we found a building where the maintenance engineer had bricked up the fresh air vents to save energy. In Washington state, one NIOSH investigator of a sick building found heavy-duty polyethylene sheets sealing off the fresh air intakes. These had

Table 4–4
COMPLETED NIOSH INDOOR AIR QUALITY INVESTIGATIONS

Problem	*Number*	*Total*
Contamination (inside)	36	17.7
Contamination (outside)	21	10.3
Contamination (building fabric)	7	3.4
Inadequate ventilation	98	48.3
Hypersensitivity pneumonilis	6	3.0
Cigarette smoking	4	2.0
Humidity	9	4.4
Noise/illumination	2	1.0
Scabies	1	0.5
Unknown	19	9.4
Total	203	

Source: James Meluis, Kenneth Wallingford, Richard Keenlyside, and James Carpenter, "Indoor Air Quality-The NIOSH Experience." Paper presented at meeting of the American Society of Heating, Refrigeration, and Air Conditioning Engineers. Atlanta, Ga., April 1985.

been installed two years earlier to reduce the levels of silica dust being carried into the building from Mount St. Helens. There are also numerous incidences of inadequate ventilation due to hidden blockages inside ducts. Using fiber-optic technology, we have found many classic examples, where turning vanes, dampers, and reheat coils inside ducts have been totally sealed with massive accumulations of dirt, loose insulation, and similar other problems.

Perhaps the most serious problem of ventilation is that there is no effective legislation mandating the uniform use of minimum fresh air requirements. If authorities could agree on a specific design code, it would be relatively easy to enforce adherence to such codes during design and construction. However, a major problem after the design and construction phase is that there is neither a legislative structure nor a practical policing methodology to ensure that the operators of buildings run ventilation systems according to such designs.

Effect of Energy Conservation

Some of these examples of inadequate ventilation were due to ignorance or accidents. However, the complex of symptoms that I have mentioned—the sick building syndrome—may result primarily from energy conservation efforts to seal buildings and reduce the infiltration and exfiltration of air. Such efforts have reduced the natural infiltration of fresh air that previously existed in many buildings, exacerbating the often undiscovered problem of a poorly designed or maintained heating, ventilation, and air-conditioning system.

In addition to tightening buildings and sealing windows, building managers have shut down air-conditioning systems at night and on weekends in an effort to lower energy costs. When the air-conditioning is shut down in humid climates, condensation builds up, and water settles inside the ductwork. If dirt is present in damp ductwork, spores and microbes can flourish; they spread throughout the building once the system is turned on the next morning. This often results in Monday morning complaints of building odors or building sickness that disappear during the week, only to recur the following Monday morning. To save more energy, automatic temperature controllers are used to cycle fans on and off during the day. Vibrations from the start-up of these fans can dislodge dirt and microbes trapped inside ductwork and carry them into occupied areas.

Another energy conservation effort that may contribute to sick building syndrome is the recirculation of indoor air, at the expense of fresh outdoor air. The 35 percent of the buildings mentioned were saving energy by shutting off all fresh air. In fact, 64 percent of all the buildings investigated by ACVA to date were operating with an inadequate level of ventilation.

Extremely bad distribution of air throughout the building is common, especially in systems using multiples of fan coil units mounted throughout the various floors of the building. Local thermostats switch off individual units independently of others, and microenvironments are set up. Often it is necessary to ensure that when the heating or cooling is not required, all the fans should be left running to aid circulation throughout the areas concerned.

Variable air volume systems (VAV) using VAV mixing boxes mounted in the ceiling void frequently have louvers opening into the void. When certain temperature conditions are met, the louvers open, and return or exhaust air from the void can be induced into the supply air, bypassing the filtration system. We have found fibrous glass, asbestos, fungi, and tobacco smoke to be recycled throughout an office due to this design.

More and more frequently one finds supply fans automatically throttled back for energy savings, say to 25 percent of their rated capacity. If the exhaust fan is not adjusted at the same rate, the exhaust fan can overpower the supply fan, and no fresh air gets into the building. The open fresh air louvers now act as additional exhausts, and the whole building runs at negative pressure. When this occurs, unfiltered outside air infiltrates into the building or, worse, exhaust fumes are sucked up from underground garages.

In addition, the substitution of low-cost, low-efficiency filters to reduce pressure drops and save energy seriously reduces the efficiency of building filtration systems and can lead to serious indoor air quality problems.

Ventilation Costs

Without doubt, the major resistance to increasing ventilation rates has been the cost of such increases. Most companies have incorporated energy management problems and new operating budgets based on saving every energy dollar possible. In an address to the Association of Professional Energy Managers in 1986, we pointed out that since their very salaries and bonuses are dependent on reduced costs, it would be an anathema for them to consider increasing energy usage and cost by increasing ventilation.

Forward-thinking companies, however, should look beyond the constraints of budgets of energy managers. Consider the following: the average heating, ventilation, and air-conditioning operating costs of a typical 100,000-square-foot building in the Washington, D.C., area would be $50,000 per annum. A commendable target for energy saving by shaving on ventilation may be, say, 25 percent savings, giving a useful $12,500 per annum. Many buildings are many times larger than 100,000 square feet, so these savings are an attractive goal.

Now consider the payroll costs for people in these buildings. Using typical averages, there are 150 square feet of space per employee; therefore, each 100,000 square feet would support 667 people. Supposing we paid each staff member only $15,000 per annum for the salary plus payroll costs; the salary bill (667 × $15,000) would be approximately $10 million per annum per 100,000 square feet. Thus, each 1 percent absenteeism costs $100,000 per annum. Typical absentee rates run at 3 to 7 percent, and 33 to 50 percent of all absenteeism is estimated to be due to upper respiratory problems, many of these due, no doubt, to dusts, bacteria, fungi, fibers, chemicals, smoke, carbon monoxide, oxides of nitrogen, and so on.

In short, what does it profit a company to save $12,500 in energy savings if that small saving causes potentially hundreds of thousands of dollars in absenteeism, not to mention lost worker efficiency? It is small wonder that some European countries, including Denmark, West Germany, and Switzerland, have introduced legislation mandating that steps be taken to prevent the buildup of internal pollutants. The United States is destined to follow that course either by slow evolution or legislation precipitated as a result of court actions brought by individuals or by trade unions making the building owners, architects, designers, and operators responsible for the health and welfare of their staff or tenants.

The Solution to Indoor Pollution

The steps to better-quality, cleaner, and safer indoor air are relatively simple and logical. We must first recognize the root cause of the problem instead of reacting to the symptoms. For example, tobacco smoke accumulation in a building is a symptom of bad ventilation. Banning smoking without addressing poor ventilation is no long-term solution. All the invisible and potentially far more dangerous indoor pollutants will still be present in that poorly ventilated environment.

The so-called clean indoor air acts so prevalent in diverse cities, townships, and states are a political sham unless they address the fundamental problems of ventilation, filtration and hygiene.

Seducing the general public to ban smoking by having them vote for clean indoor air without addressing the ventilation rates and filtration standards leaves that same public in a fool's paradise. Indoor air pollution can still be rampant even where it is invisible. We cannot see carbon dioxide or monoxide, formaldehyde, ozone, ammonia, airborne fibers, bacteria, or fungi. Since these pollutants make us sick and can even kill us, a clean indoor air act cannot address these problems.

The correct solution is an elementary application of common sense. The three major common denominators of indoor pollution are poor ventilation, inadequate filtration, and lack of hygiene. Any effective clean indoor air act should address these three areas. We should stipulate minimum ventilation rates for all buildings. Laws should enforce the application of ventilation rates in the design and ongoing operation of all buildings. No one should have the freedom to subject staff, tenants, or visitors of buildings to stale, recycled polluted air. Filtration standards should be defined and programs introduced to ensure that air filters are well installed, regularly inspected, and, when necessary, changed. Finally, ventilation systems and their associated ductwork, the very lungs of the buildings, should be routinely examined, and if contamination is present, it must be removed.

With these three simple steps, the air quality throughout every building will be dramatically improved. All indoor pollutants, gases, vapors, particulates, and microorganisms will be reduced. Absentee rates will drop, worker productivity will increase, staff morale will improve, and company profits will rise. Surely these are worthwhile returns for the small investment necessary to diagnose and rectify the true causes of indoor pollution.

5

Cigarettes and Property Rights

Walter E. Williams

The kind of man who demands that government enforce his ideas is always the kind whose ideas are idiotic.

—Henry Louis Mencken

WAITING in a long reservation line for a flight from San Francisco to Philadelphia, I lit a cigarette. After a few puffs, a middle-aged woman in front of me said, "Sir, would you mind putting that cigarette out?" Being bored and willing to converse, I asked, "Do you own the air?" "What does that have to do with putting out that cigarette which is causing a nuisance?" she replied. "It has a lot to do with it," I admonished. Little did she know that she was about to receive a minilesson on property rights. The complex property rights issue is one ignored by both smokers and antismokers.

Antitobacco campaigns have waxed and waned for several centuries. King James I, who saw smoking as a pernicious habit borrowed from wild, uncivilized Indians, was among the first prohibitionists. French Cardinal Richelieu objected to smoking and thought a tobacco tax would not only check its spread but be a great source of government revenue as well. During the Napoleonic period, Vienna, Paris, Berlin, and parts of Switzerland passed laws prohibiting smoking in the streets. Queen Victoria hated smoking; she made royal guests at Windsor Castle smoke into fireplaces so the smoke could go out the chimney.

In the United States, Lucy Page Gasten led the first antitobacco movement around the time the Women's Christian Temperance Union movement was campaigning against alcohol consumption. Her movement was followed by Dr. Charles Pease of the Non-Smokers' Protection League of America. Henry Ford, the automobile industrialist, was a fanatic antismoker who tried in vain to get his employees to stop smoking. The early antitobacco campaign had modest success. Twelve states had statutes that either banned or restricted the sale or use of cigarettes; however, along with the demise of the Eighteenth Amendment, these statutes were repealed.

Early antitobacco campaigns centered their attention on how cigarette smoking harms the smoker. Experts of the time alleged that smoking was not only harmful to physical and mental health but led to a breakdown in morals, leading first to alcoholism and then to opium addiction. Moreover, they said, adult smoking would lead increasing numbers of children and young adults to take up the habit. Their message and efforts fell on deaf public ears as cigarette smoking grew in America.[1] But early medical experts were not in full agreement. Cornelius Bontekoe, a celebrated Dutch physician, said, "There is nothing so good, nothing so estimable, so profitable and necessary for life and health as the smoke of tobacco, that royal plant that kings themselves are not ashamed to smoke."

Smoking Harms the Smoker

At the outset, let me be clear. Although I possess no special expertise, I accept the medical evidence that cigarette smoking is a major contributor to several health disorders, such as lung cancer, heart disease, and respiratory distress. On the other hand, not everyone who smokes suffers from these disorders. Whether an activity such as cigarette smoking is harmful to the individual is not the relevant criterion for law in a free society.

Many human activities can result in personal harm. Skiing, football, boxing, skydiving, and swimming are activities where a person puts safety at risk. Soft drink consumption by children may lead to hyperactivity and aggravate allergies. Medical experts agree there is no dietary reason for adding salt to foods. Yet many people do so and subsequently increase their risk of hypertension and other circulatory diseases. Sunbathing often leads to skin cancer. A higher risk of coronary disease is associated with high cholesterol consumption. Automobile driving, mining, and bridge construction often result in injuries and fatalities. These are but a few ways we take risks with the length and quality of our own lives. Yet no one, I suspect, would advocate legal prohibitions against sunbathing, salt, red meat consumption, and other harmful and risky activities. Instead we simply leave it up to education and persuasion but otherwise leave the individual free to choose.

In a free society, each individual owns himself or herself. That means and requires that each person be permitted to take chances with his or her own life. If this person were to be prevented from engaging in every behavior that risked health and safety, he or she would not be free. He or she would be like a caged canary—safe but a prisoner. I have posed the following question to class after class over twenty years: Would you be satisfied if government had the power to prevent you from doing everything that might cause you harm? I have yet to receive an affirmative answer.

There is the frequent assertion that people who place themselves at risk by smoking, riding a motorcycle without a helmet, or engaging in dangerous activities may become medically indigent and hence a financial burden to society. The conclusion reached by this observation is that since society has the ultimate responsibility of paying the bill, it has a right to restrict people's right to take risks with their health and safety. Such a position raises an important question: if intervention, for this reason, is deemed legitimate in the case of cigarette smoking or motorcycling, is it not a logical extension of the same principle to regulate other activities, such as sunbathing and salt consumption, that may jeopardize health and safety and cause people to become a financial burden on society?

The argument that if people are permitted to take chances with their life and good health may lead to circumstances where they become a financial burden to society is not a good one against individual freedom. Instead it is a better argument against socialism. In a free society, when an individual harms himself or herself and is thereby unable to make a contribution to the productive output of society, he or she is penalized by having no wages and hence no ability to make claims on goods and services produced by society unless he or she had foresight to purchase disability insurance, had caring family or friends, or was lucky enough to receive charity. In a free society, individuals bear most of the cost of harming themselves, or the people who bear the cost do so voluntarily. Who bears the cost of mistakes is one way to distinguish between free markets and socialism. In free markets, costs of mistakes are borne by the individual who makes them. Under socialism, costs of people's mistakes are politically spread among the society at large.

Having failed in their efforts to persuade people to give up cigarette smoking voluntarily for health reasons, the antismoking campaign has focused its attention on the claim that others, being in the presence of smokers, or what they call involuntary smoking, risk adverse health effects. Antismokers now concede that cigarette smokers have the right to harm themselves but not others. The medical evidence on the effects of "secondary" cigarette smoke is less clear, and there is considerable controversy.[2] But it seems that some people with allergies, asthma, bronchitis, and other respiratory problems suffer in the presence of cigarette smoke. Moreover, just the odor of cigarette smoke is highly offensive to many people. These factors do not make cigarette smoke unique. Some people suffer the same allergic and respiratory distress in the presence of fumes or vapors from household cleaners, hair sprays, perfumes, deodorants, after-shave lotions, and other toiletry items.

Harm to Others

Harmful or beneficial side effects may arise in the course of consuming some commodities and affect people not directly involved in the consumption of the good in question. Economists refer to these side effects as externalities. External economies of consumption occur when one person's consumption indirectly benefits another. When one home owner makes increased expenditures to maintain property, he or she unavoidably raises the attractiveness and hence the value of neighbors' property. A person wearing a lotion pleasing to others unavoidably benefits them. External diseconomies of consumption occur when one person's consumption indirectly harms another, as in the case of an airplane flying over a property or a person dressed in displeasing attire. In either externality case, there is no compensation for the harm or benefit.

When external diseconomies of production or consumption occur where A's activities harm B, it is intellectually shallow to ask how we can stop A from harming B. The reason is that externalities are always reciprocal. That is, to prevent B from being harmed automatically harms A. For example, a motorcyclist riding down an otherwise quiet road momentarily disturbs, and hence harms, the residents. Shortsighted inquiry would ask how we can prevent the cyclist from harming the residents. But such a question fails to evaluate and consider the harm that can come to the motorcyclist by preventing him or her from using the road.

We can think of other nuisances, such as airplanes flying over congested areas, traffic noises, or schools adjacent to residential areas. In each case a home owner may be harmed by the consumptive activities of others; however, to prevent the nuisance imposes a harm on those participating in those consumptive activities. Although superficial analysis might suggest that home owners are harmed by these nuisances, it is not always clear they are harmed on balance. The presence of neighborhood schools may raise property values. The presence of a nearby airport may stimulate economic activity and in turn raise property values of the people annoyed. In other words, the same externality may simultaneously produce both harmful and beneficial effects.

It is difficult to conceive of a world where all harm has been outlawed. In such a world, activities we normally take for granted could not occur. Marriage is one of them. When a man or a woman marries, he or she unavoidably harms other people. In other words, he takes a woman who might have been the apple of some other man's eye, and she takes the man who might have been the apple of some other woman's eye. The more attractive the marriage partners are, the more widespread is the harm in terms of the prospective mates they had to say no to.

The same reciprocal externalities apply to the smoking controversy. Some nonsmokers are harmed by smokers. Public debate centers around ways to prevent smokers from causing harm to nonsmokers. This is intellectually shallow because it ignores the potential harm done to smokers. The harm done to smokers is that of denying them the opportunity to engage in an activity they enjoy.

In the case of externalities, there is no way of unambiguously assessing which harm is more devastating or more important to prevent. To attempt to do so is to make interpersonal utility comparisons that economic theory proves impossible. In other words, there is no way to establish whether having a noisy school in a neighborhood is intrinsically more, or less, valuable than having quiet in the neighborhood. It is a matter of personal values. Similarly, there is no way to assess whether the harm done to a nonsmoker is more important than the benefit gained by the smoker or whether the gain to the nonsmoker through smoking bans is more important than losses to the smoker.

Moreover, the nonsmoker, as well as the smoker, places subjective values and trade-offs on the harm or benefit he or she is willing to accept. Consider the following: Would all nonsmokers turn down an invitation to a White House dinner knowing they would be seated at the table with a president who smokes? Would a nonsmoker faced with a disabled car refuse the offer of a passing motorist who smoked? Would a nonsmoker turn down an interview for a job he has always wanted because the employer smoked?

The Right to Clean Air

Some people in the antismoking movement have argued that a basic unenumerated constitutional right is the right to breathe clean air. Taken at face value, such a statement would produce ludicrous results. Pollution of one form or another is the by-product of all consumption or production. The production of dresses, soap, medicine, and everything else uses energy. A by-product of energy consumption is pollution. The consumption of auto services, use of hydrocarbons, and even the act of life itself (taking in air and expelling carbon dioxide) produces pollution. Hence, what is the meaning of the statement that people have a constitutional right to clean air? Obviously anyone who makes such an assertion must be prepared to offer a trail of modifiers or call for the absolute curtailment of production and consumption.

At least one court has recognized this in *Gasper* v. *Louisiana Stadium and Exposition District* (577 F.2d 897 [1978]). The court disagreed with the right to clean air argument saying, "To hold that the First, Fifth, Ninth, or Fourteenth Amendment recognize as fundamental the right to be free from cigarette smoke would be to mock the lofty purposes of such amendments and broaden their prenumbral protections to unheard-of boundaries."

Property Rights

At the core of the smoking controversy lies the property rights issue. Property rights have to do with who makes decisions to acquire, use, and dispose of resources. Property rights are said to be held privately when the person deemed the owner has (1) exclusive rights to use, (2) rights to receive all benefits and duty to pay all costs, and (3) rights to transfer these property rights (sell).

When property rights are well defined and transactions costs (costs associated with information and contract enforcement) are zero, external costs and benefits are internalized regardless of the property rights assignment. In other words, the person whose consumptive activity confers an indirect benefit on another is compensated, and the person whose consumptive activity imposes an indirect cost on another compensates that person. If I break a person's window while playing baseball, I compensate the owner; if I collide into your car, I compensate you for the damages. Alternatively, if I confer a benefit through mowing another's lawn, which imposes a cost on me through sacrificed leisure, I am compensated by receiving wages. In the case of well-defined property rights, there are no externalities.

Smoking in Homes

The common law presumption is that the decision whether smoking is allowed in private homes is left in the hands of the person(s) deemed the owner(s). This means, regardless of the external effects, that nonsmokers have the right to restrict smoking in their homes but not in somebody else's house. I suspect even the most zealous antismoker would be offended by a law banning smoking in private homes even though smoking in private homes may pose the same risks to the smoker and those around him or her, such as family members and guests, as it would on airlines and in restaurants.

I imagine the counsel to a person who complained about being in the presence of smokers at another's house would be a suggestion to leave the premises. Similarly, a person who complained about others smoking in his house would receive similar counsel: tell the smoking guests to refrain or leave. Resolution of conflict over smoking is straightforward and reasonably simple where clearly defined property rights prevail. I imagine a defendant in a tort case in which the plaintiff claims injury by cigarette smoke would use the affirmative defense of assumption of risks; the defendant would claim that the plaintiff had knowledge of a condition or situation that was dangerous and yet exposed himself or herself to the hazard created by the defendant.

Smoking in Publicly Owned Places

The conflict of wishes between smokers and nonsmokers becomes a more contentious issue in publicly owned places because property rights are ill defined in the sense of who owns what. Publicly owned places include municipal airports, parks, schools, and state and federal buildings. Presumably smokers as well as nonsmokers pay taxes, and both can claim partial ownership of publicly owned places. There is nothing in principle that says whose rights should prevail: the right to smoke and have cigarette-polluted air or the right to ban smoking and have nonpolluted air in publicly owned places. More often than not, whose wishes prevail is a result of which group has the dominant political power at the moment.

Smoking in Privately Owned Establishments

Smoking in factories, offices, restaurants, airlines, and department stores, which cater to the public as workers or customers, are the current focus of the antismoking movement. These establishments are clearly privately owned. The fact that they do business with the general public does not make them publicly owned. In their cases, private property rights are well defined and should resolve the smoking controversy just as it does in the case of privately owned homes. The owner makes the decision whether smoking is to be permitted on the premises and accepts the economic consequences.

What are the options available to the nonsmoker or antismoker worker under these circumstances? The most obvious is for the worker to withhold his or her productive services from employers who permit smoking on their work sites. Another option is to try to convince, cajole, or embarrass smoking coworkers into abstention.

Employers have a common law duty, given the state of technology, to ensure a safe workplace. There is a similar common law expectation that home owners must take reasonable precautions to ensure safety for their visitors. At the same time the common law holds employers and home owners accountable to take reasonable measures to ensure the safety of employees and visitors, it also places a burden on the latter to take reasonable measures to protect themselves.

Unlike the cases of slippery stairs, radioactivity, and certain noxious fumes, the fact that coworkers and others are smoking is visible, apparent, and cheap to detect. Therefore a person who might suffer injuries from being in the presence of smokers knowingly assumes the risk of harm. The remedies available to a worker in a workplace where the employer permits smoking are identical to those available to a nonsmoker who visits a house where the owner permits smoking: avoid the danger by leaving. To rejoin to this remedy by saying the visitor has choices while the worker does not is the same as asserting the intersection of two improbable events: there are no other jobs available, and all employers permit smoking.

Obviously similar remedies are available to nonsmokers in choice of restaurants, bars, and other places of entertainment. They are free to withhold patronage from establishments that permit smoking. To facilitate nonsmokers' options or, for that matter, smokers' as well, all that is necessary is for owners to make readily available information about whether smoking is permitted at their establishments.

Abrogation of Private Property Rights

Under private property rights, and implications for resource usage, some people may bear costs that can be avoided with government allocation of resources. One experience can demonstrate this principle. The hotel bellman summoned a taxi for me. Upon entering the taxi, I noticed a sign saying, "No Smoking." I told the driver that since he did not permit smoking, I would get another taxi. The driver used abusive language to exhibit considerable displeasure when I got into another taxi.

The driver's displeasure might have resulted from any one of several factors. First, he had to get out of the taxi waiting line for naught. Second, he lost a fare and had to wait for another. Third, he might have objected to my rejecting his services solely because of his prohibition of smoking in his vehicle. There were some costs borne by me as well, namely those associated with finding another taxi. In this instance, they were low since there was a taxi queue.

The taxi driver would have not encountered the inconvenience and lost earnings if there had been a law against smoking in taxis because there would have been no alternative to a no-smoking taxi. On the other hand, if there were a law mandating the right to smoke in taxis, the experience would have been cheaper for me; I would not have to search for another taxi.

One of us would have gained from the abrogation of the property rights of the other. Abrogation of property rights has become the most popular route in the antismokers' drive for prohibition. They have realized the age-old practice of using the legislature to accomplish what they deem more costly through voluntary market transactions. Their agenda is to use the legislature to take away part of the owner's decision on how his or her property is used.

A similar scenario emerged in California. In April 1986 the Beverly Hills city council enacted a ban on smoking in restaurants. By August, Beverly Hills restaurants had lost 30 percent of their customers to restaurants in nearby towns, which allowed smoking. The city council then adopted new rules, permitting smoking in restaurants provided there were no-smoking areas and adequate ventilation.[3]

The problem for the Beverly Hills restaurants was that the city council had the power to ban smoking in Beverly Hills but not elsewhere. Moreover, there was a loophole in the law exempting bars and cocktail lounges from the smoking ban, and some Beverly Hills restaurants "became" lounges. The combined effect was losses to no-smoking restaurants resulting from customers taking their business elsewhere.

Beverly Hills restaurateurs would have benefited from a similar ordinance abrogating the property rights of restaurateurs in adjacent towns. Broader smoking restrictions would have reduced their revenue losses by denying customers alternatives. This suggests that if businesspeople perceive a high probability of enactment of smoking bans, they can be expected to join the lobby for the enactment of the ban. This is true whether or not individual businesspeople are against it. It is in their financial interests to make the application of the smoking ban as broad as possible so they do not experience losses like those experienced by the Beverly Hills restaurants. Therefore, we would not expect one private airline to enact smoking bans on its aircraft, even if its owners were antismokers; they would fear the revenue loss. We would expect them to seek legislation outlawing smoking on aircraft altogether so as to avoid revenue losses from customers "voting on their feet."

Use of the State for Windfall Gain

People often seek to use state powers to redistribute wealth from one class of citizens to another. An example from housing will be used to illustrate the process. Consider the person who purchases a house for $100,000 in a neighborhood where identical houses across the road sell for $125,000. The reason why houses on his side of the road sell at a discount is because a railroad runs behind the properties, causing noise and vibration. For this nuisance, home owners with property adjacent to the railroad have been compensated, through the market, by a property discount of $25,000.

The home owners on the noisy side of the road might form a political coalition and appeal to the legislature to enact a law forcing the railroad to relocate its tracks or invest resources in noise- and vibration-abatement equipment. They might buttress their complaint by bringing scientific evidence that the train nuisance harms their children by having their sleep disturbed by passing trains.

If the state legislature ruled in favor of the home owners, the railroad company would have to incur the costs of noise abatement or track relocation. As a result, not only would home owners now have undisturbed sleep, they would receive windfall gains in the form of a wealth transfer from the railroad. Their homes, which differed from their across-the-road neighbors only by the nuisance of passing trains, would begin to appreciate in price. Assuming the train nuisance was the only difference, after its removal, their houses would sell for $125,000, the price of other houses not encumbered by the railroad nuisance.

Obviously similar use of state power can be employed in many other ways to confer windfall gains at the expense of windfall losses to others who are not as politically strong. This principle can be applied to the issue of smoking bans in the workplace. A nonsmoker might take employment in a business establishment knowing the employer permits smoking on the premises. Rather than not take the job in the first place or quitting if the smoke becomes intolerable, he or she can use the legislature to abrogate property rights of the owner by outlawing cigarette smoking. The nonsmoker's windfall gains are not monetary but are nonetheless state-acquired gains resulting from wealth redistribution that raises his or her sense of well-being at the expense of another.

The Market and Conflict Resolution

People exhibit many likes and dislikes. Often these likes and dislikes lead to conflict. Political allocation of resources raises the potential for conflict because one person benefits only at the expense of another. Political allocation of resources is a zero-sum game where the gain to one person necessarily comes as a loss to another. Market allocation reduces the potential for conflict because it is a process where all parties to the exchange benefit, a positive-sum game.

Smoking may be an annoyance and possibly harmful to those around the smoker. Such an observation does not place smoking in a category by itself. Loud music, featured at discotheques and bars, is said to lead to hearing loss and headaches. Typically the person who is offended by the loud music either leaves the establishment or does not enter in the first place. As long as loud discotheque music is contained within the establishment, most people offended by it would not descend on the state legislators lobbying for enactment of a law specifying a particular decibel range. There is no ugly confrontation. People who like loud music go to discos, and those who do not go elsewhere. It appears as though a similar response is available to people who find smoking distasteful.

Efficiency Gains through the Market

Most human conduct is regulated not by law but by force of social, custom, etiquette, and ostracism. Behavior regulated in this way includes, or included, men giving up their seat for women, boisterous behavior, child rearing, foul body odor, coughing around others, and dress codes. Violation of these customs is met with social sanctions leading to general compliance without their being regulated by threat of state violence or punishment. Customs and etiquette are ways we internalize externalities in the presence of nonzero transactions cost.

A reasonable question is whether efficiency gains can be realized using social norms, as opposed to the threat of state violence, as a regulator of annoyances related to smoking. Clearly one source of efficiency is the case where nonsmokers have neutral preferences about being around smokers. If there is a law prohibiting smoking, it is more likely that some people will be worse off without the benefit of anybody else being better off.

Such a principle is a subject of considerable study in the area of economics known as welfare economics. Welfare economics studies the conditions under which society is made better or worse off. The maximization of society's well-being requires the optimal allocation of resources in production, consumption, and exchange.

The efficiency rule, a cornerstone of modern economics, is known as Pareto optimality named after Vilfredo Pareto, its turn-of-the-century Italian discoverer. According to Pareto-optimality criteria, any change that improves the well-being of some individuals (in their own estimation) without reducing the well-being of other individuals (in their own estimation) clearly improves the welfare of the group as a whole. Such a change will move the group from a Pareto-nonoptimal position to a Pareto-optimal position.

Applying Pareto criteria to the smoking controversy, if a set of conditions prevails where some people will derive increased enjoyment from smoking and others around them are not made worse off (in their own estimation), then the group is made better off if smoking is permitted. Under such circumstances, antismoking laws serve no purpose other than preventing greater enjoyment for the group. Etiquette, in the form of "Do you mind if I smoke?" would be a better regulator.

Common Law and Smoking

In traditional common law of torts, there appears to be no actionable remedy for a plaintiff bringing a nuisance or injury case against a home owner, restaurateur, café owner, or manufacture owner defendant claiming injury by the secondary effects of cigarette smoke in the defendant's establishment. It would appear that the defendant would have several common law doctrines for defense.

First, though controversial, is the coming-to-the-nuisance doctrine cited in *Bove v. Donner-Hanna Coke Corporation* where the court rejected the plaintiff's request for an injunction saying:

> With all the dirt, smoke and gas which necessarily come from factory chimneys, trains and boats, and with full knowledge that this region was especially adapted for industrial rather than residential purposes, and that factories would increase in the future, plaintiff selected this locality as the site of her future home. She voluntarily moved into this district, fully aware of the fact that the atmosphere would be contaminated by dirt, gas and foul odors and she could not hope to find in this locality the pure air of a strictly residential zone. She evidently saw certain advantages in living in this congested center.[4]

There is considerable equivocation on the issue of coming to the nuisance. In the *Restatement of Torts* on nuisance provisions, it is argued that assumption of the risks should be a defense in nuisance actions to the same extent it is a defense in other tort actions. Also, in coming to the nuisance, *Restatement of Torts* holds that "the fact that a plaintiff has acquired or improved his land after a nuisance interfering with it has come into existence is not in itself sufficient to bar his action, but is a factor to be considered in determining whether the nuisance is actionable."[5]

Another common law doctrine almost wholly ignored in nuisance law is the duty to mitigate damages. In some cases, the parties to a conflict can avoid or reduce damages leading to litigation by taking reasonable low-cost steps. For example, if a nonsmoker sees or knows that the owner of a private establishment permits smoking, which the nonsmoker perceives as hazardous, he or she may simply leave or not enter the premises.[6] This duty to mitigate damages seems to be related to court decisions where people come to the nuisance and are thereby held to have assumed the attendant risks. None of this argument suggests that property owners do not have a duty to ensure the safety of their premises. But it would appear that the safety obligation applies to hazards that cannot be reasonably known or anticipated by others entering the premises.

The tradition of common law respect for private property may explain why people who wish to abrogate the private property rights of others choose statutory relief rather than appeal to civil courts.

Conclusion

Cigarette smoking is a private act that produces external diseconomies for some people. For that reason antismokers call for government regulation. However, it is possible to prove that every private act has external effects of one sort or another. The list of private acts that can reduce the physical and psychological well-being of other people is enormous and includes loud music, rambunctious behavior, poor table manners, barking dogs, disheveled appearance, sexual promiscuity, obscene language, trespass, obesity, meat eating and flatulence. Therefore to call government regulation of private acts solely because they produce external costs and reduce the welfare of other people is to call for government regulation of everything.

Social costs or external diseconomies are a politician's dream. It gives them what they consider the moral justification for having government play a larger and larger regulatory role in our lives. For political gain, it gives them power to grant favors and privileges to one group of citizens at the expense of another group of citizens. The smoking controversy is no exception to the ongoing efforts of politicians, promoting class interests, to grab power. While a case might be made for political regulation of smoking in publicly owned places, no case can be made, and simultaneously respect private property rights, for political intervention to regulate smoking on private property.

Notes

1. See S. Wagner, *Cigarette Country: Tobacco in American History and Politics* (New York: Praeger, 1971).
2. We must always be alert to people using science as a means to accomplish hidden agendas. See Edith Efrom, *The Apocalyptics: Cancer and the Big Lie* (New York: Simon & Schuster, 1984), chaps. 1, 2.
3. *Time*, August 3, 1986, p. 23.
4. Cited by Charles O. Gregory, Harry Kalven, Jr., and Richard A. Epstein, *Cases and Material on Torts,* 3d ed. (Boston: Little, Brown, 1977), p. 535.
5. Ibid. p. 535.
6. See Bruce Ackerman, ed., *Economic Foundations of Property Law* (Boston: Little, Brown, 1975), pp. 293–294.

Bibliography

Ackerman, Bruce, ed. *Economic Foundations of Property Law.* Boston: Little, Brown, 1975.
Chung, Steven N.S. *The Myth of Social Cost.* Menlo Park, California: Cato Institute, 1980.
Count Corti. *A History of Smoking.* London: George G. Harrop & Co., Ltd., 1931.
Gregory, Charles O., et al. *Cases and Material on Torts.* 3d ed. Boston: Little, Brown, 1977.
Trayer, Ronald, and Gerald Marble. *Cigarettes and the Battle over Smoking.* New Brunswick, N.J.: Rutgers University Press, 1983.
Wagner, S. *Cigarette Country: Tobacco in American History and Politics.* New York: Praeger, 1971.

6

Environmental Tobacco Smoke

A Smoke Screen for Management?

René Rondou

IMAGINE that you are a union shop steward in a factory somewhere in industrial America. The management of your factory is not the most appreciative of union activities; in fact, like many other companies in the Reagan eighties, it barely tolerates your union and your enthusiastic performance of your steward's duties. One day you come to work to find that one aspect of your personal behavior has been unilaterally outlawed in your workplace. If you persist in this behavior, you can be demoted, transferred, or fired, and management will put you through periodic medical examinations and random inspections of your work station to ensure compliance with its policy. You know that these regulations will be enforced selectively and that, as a thorn in management's side, you are particularly vulnerable to a job action against you. But you have no choice but to submit to management's directive. Like 30 percent of all adult Americans, you are a smoker, and your decision to smoke has made you subject to employment restrictions over which you and your union have had no voice.

This scenario is not fiction. It is happening all across the United States as employers seize on antismoking hysteria as a reason for imposing the strictest restrictions upon the personal choices of their employees. It's happening not just in factories to blue-collar workers. More and more white-collar workers in shops, office buildings, schools, and hospitals have found themselves subject to the arbitrary imposition of antismoking policies by their employers, with the penalties for disobedience ranging from forced attendance at smoking-cessation clinics to summary loss of their jobs.

The suspension of traditional personal freedoms in these circumstances is obvious, as is the threat to the long-standing right of unionized workers to bargain collectively with management about workplace rules. The worst threat represented by arbitrary antismoking regulations imposed by management is more insidious and also more dangerous. It has become clear that antismoking regulations can act as a cover for management, disguising major threats to workers' health and safety from poor indoor air quality. The transfer of responsibility for indoor air quality and safety from management to the mass of workers has been done silently, deviously, and to some extent with the approval of the U.S. government.

(Good) Air Isn't Free

Since the onset of the Industrial Revolution, the link between workers' health and the quality of the air they breathe on the job has been well recognized. Traditionally heavy industry, with its smoky factories and toxic substances, was the acknowledged source of respiratory ailments, skin diseases, contagious viruses, and cancer. Many of this country's unions were founded in response to these threats to industrial workers' health and safety. Most people assume that organized labor has succeeded in its demands for safe factories, equipment, and procedures, especially since it became the federal government's regulatory mandate to protect workers from occupational diseases.

Although most industrial environments have become cleaner and better ventilated in the last fifty years, the invention and use of new and toxic chemicals and manufacturing procedures has created different and more subtle threats to the health of factory workers. Meanwhile, white-collar and clerical workers performing their duties in supposedly clean office buildings have also found themselves at risk from invisible pollutants circulating in the hermetically sealed air. The combination of new technologies and new methods of building means that today's workers are as much at risk from invisible dangers as their grandparents were from unsafe equipment and firetrap factories.

These dangers are confirmed by statistics. Every year an estimated 100,000 workers die of occupation-related diseases. An estimated 1 million more are disabled from exposure to dangerous substances. About 11 million workers are exposed to known carcinogens by their employers; some estimate that between 20 and 30 percent of all cancers in the United States are occupational in origin. These exposed workers did not choose their risk before they chose their livelihoods; many, if not most, of the most serious conditions are a direct result of employers' refusal to protect their employees through better plant engineering, improved ventilation, and safer manufacturing processes.

While the problems of indoor environments may have changed, the responsibility of management to provide safe workplaces has not. The refusal of many companies to confront the need for better construction and maintenance of their plants and office buildings has found a convenient camouflage in antismoking rhetoric. By placing the burden of proof on the employee's personal life-style, management abdicates its responsibility to make fundamental (and sometimes expensive) changes in the work environment and procedures in the name of a pious concern over workers' individual behavior. It is far easier—and cheaper—to put smoking at the top of the priority list for workplace safety and ignore other employer-directed claims on indoor environmental protection.

The Facts on Smoking and Ventilation

The specific hazards of indoor air pollution are demonstrated by recent history. In 1968, a flulike epidemic struck ninety-five out of one hundred workers at the health department building in Pontiac, Michigan. The source of the illness was traced to a defective central air-conditioning system. In 1976, twenty-nine American Legionnaires died, and another 153 became ill, from an airborne bacterial infection since named for its victims. And, with great irony, a quarter of the workers in one area of the Health and Human Services building in Washington, D.C., contracted hypersensitivity pneumonia, a lung condition traced to a ceiling leak of greasy water from a malfunctioning dishwasher.

These are the most extreme examples of what has come to be known as a "tight" or "sick" building syndrome. Often the newest and most glamorous office buildings—the kind where windows cannot be opened and a computer controls the heating, ventilating, and air-conditioning (HVAC) systems—are the most dangerous to workers. The energy crisis of the 1970s and the resulting concern by building owners and operators over cost-effective energy use has made it more economical, and thus attractive, to reduce the flow of fresh air into the building. The resulting poor ventilation creates a situation where tobacco smoke is the only visible symptom of a much graver problem.

According to the National Institute for Occupational Safety and Health (NIOSH), only 2 percent (4 out of 203) of buildings inspected following occupant complaints about indoor air quality could trace these complaints to environmental tobacco smoke. Almost half the complaints—typically headaches, runny noses, dizziness, and chronic respiratory illnesses—were traced to inadequate ventilation. In another 18 percent of the buildings, poor indoor air was caused by circulating contaminants, such as chemicals from duplicating machines, insecticides, and sulfur dioxide from heating systems. In another ten percent of the buildings, other contaminants, including carbon monoxide from basement parking garages, caused the problems.

The conclusion that can be drawn from the NIOSH study is that tobacco smoke is in itself not the source of poor indoor air conditions. Rather, as the only visible air contaminant, its lingering presence is an indication of a sick building, which is propagating other, more dangerous but invisible contaminants through overloaded ventilation systems. And although the NIOSH study was restricted to commercial office buildings, schools, and hospitals, its implications are equally applicable to the industrial workplace, where toxic particles such as asbestos, various dusts, fumes, and other hazards pose an even greater threat to the health of industrial workers.

The contrast between the questionable problem of tobacco smoke and the actual damage done by workplace toxins is obvious in U.S. factories. Despite standards regulating asbestos exposure, which are regarded as overly liberal by most of the medical establishment, more than 30 percent of U.S. factories are not in compliance with federal regulations. And despite the outlawing by Congress of the use of polychlorinated biphenyls (PCBs) in new electrical equipment more than a decade ago, at least 20 percent of all workers working in electrical utility plants, train yards, and hydroelectric stations are still exposed to this most carcinogenic of chemical compounds. Textile workers inhaling cotton dust, nuclear power plant employees exposed to radioactive particles, and miners who are still disabled in great numbers by black lung disease are further examples of the continuing hazards of the industrial workplace.

And what is management's solution for the electrical worker exposed to carcinogenic PCBs, for the pregnant woman typing on an unscreened computer terminal, for the municipal employee with chronic bronchitis caused by his office air-conditioner? These days, it boils down to two words:

"No smoking." And even when management deigns to procure a solution to the problems its own practices or environment create, it is usually an added imposition on the worker—cumbersome and frequently ineffective protective gear that does little to protect the worker's air supply against toxic substances.

The task for labor is clear: return the primary responsibility for a safe and healthy workplace where it belongs, with the owners and managers of the American workplace.

Negotiating for a Better Work Environment

The labor movement has already acknowledged the importance of the indoor air quality debate and its implications for the collective bargaining process. The catalyst for action was the 1985 U.S. surgeon general's report on environmental tobacco smoke. This report made dubious history by ignoring the years of proof regarding the dangers of synthetic toxins and inhaled particles in the industrial workplace and by placing employee smoking at the top of the list as a cause of occupational cancers.

In 1986, the AFL-CIO Executive Council issued a statement in response to the surgeon general's report, which noted in part:

> The AFL-CIO believes that employers will attempt to use the report to shirk their responsibility to clean up the workplace and to place blame for occupational disease on workers who smoke. . . . The AFL-CIO believes that issues related to smoking on the job can best be worked out voluntarily in individual workplaces between labor and management in a manner that protects the interests and rights of all workers and not by legislative mandate.

Unions now recognize that the issue of smoking in the workplace and environmental tobacco smoke must be dealt with in the broader context of labor-management relations. Traditionally it has been the task of the federal government, through the Occupational Safety and Health Administration (OSHA), to issue and enforce regulations concerning worker health and safety. But the dismantling of that agency and the hamstringing of its operations since 1981 has forestalled an immediate regulatory solution to the complex problem of indoor air quality. Coupled with the preference of many employers for the cheapest, easiest solution to indoor air quality problems—banning or limiting employee smoking—the responsibility for ensuring workplace safety has fallen to the labor movement.

The most important tool labor is using in protecting the rights of union members is the appropriate one: the collective bargaining process. As the representative of smokers and nonsmokers alike, unions have the legal responsibility to ensure that all members' rights are protected. This is not an easy task in the face of dogmatic antismoking activism. Negotiating fair compromises for the needs of its members and ensuring that any workplace smoking regulations result from labor-management consultation has become an important item on the collective bargaining agenda.

What do unions ask for? First and foremost, they ask for acknowledgment from management that workers' life-style choices be respected to whatever degree fair and practical. But as they negotiate to make workplace smoking a collective bargaining issue, unions are also holding management responsible for the situation in which smoking is only one small aspect: the overall safety and health of the indoor working environment. Unions have begun to insist that management literally clean its own house by monitoring and improving air quality and safety before challenging the lifestyle choices of employees. This battle is complex and difficult and is being waged on several fronts.

In the industrial workplace, labor has begun by challenging the findings of the surgeon general's report. This report slavishly follows the surgeon general's own political agenda and attempts to establish employee smoking as the chief cause of occupational cancer and lung diseases. With little scientific backup and a fine disregard for contradictory medical evidence, this report gave employers the ammunition they needed to divert attention and resources from their own responsibilities to those of their workers.

Unions support management's efforts to provide counseling and help for smokers who want to stop smoking. But when these programs are initiated—and backed up by harsh, unilateral antismoking policies—as a substitute for cleaning up dangerous workplaces, it is a mockery of workers' needs. Telling a smoker to quit is cheap; removing asbestos fireproofing or old electrical transformers containing PCBs is expensive. Unions have discovered that the real challenge in the workplace smoking debate is convincing employers to strengthen the standards for exposure to work site toxins and to circumvent employers' attempts to blame tobacco smoke for occupational illnesses.

In cases where union OSHA experts conclude that workplace smoking may pose special problems, unions attempt to work with management to establish the necessary rules for governing smoking behavior. Union-oriented health and wellness programs, which include information about handling and avoiding hazardous substances, and assistance programs for exposed workers (including voluntary smoking cessation programs) are appropriate responses when complemented by a rigorous safety program, and they have been endorsed by the AFL-CIO.

The White-Collar Blues

The controversy is somewhat different for white-collar workers. While industrial sites have long been acknowledged as a source of disease-causing substances, most white-collar environments are assumed to be clean and safe by employers and employees alike. The hazards of poor ventilation and indoor and outdoor contaminants are poorly understood by most users of commercial office space. And the builders and managers of these buildings, who do understand the problems created by badly designed and maintained facilities, choose to ignore sick building symptoms in favor of lower energy and maintenance costs.

Labor's first task in this environment is the education of both management and its own membership about the dangers of workplace-originated discomfort and illness. Unions are beginning to insist that management implement comprehensive air quality investigation programs as part of any attempt to deal with smoking issues. Often such an investigation demonstrates that all the occupants are suffering from an indoor air pollution problem of which lingering tobacco smoke is only a symptom.

Both employers and employees should understand the essentials of building maintenance: that fresh air intakes are clean and open; that filters are well installed and have a minimum filtration efficiency of 40 percent; that humidifiers and condensate trays be kept clean and drained; and that regular inspection of the ventilation ductwork be undertaken (and the results publicized) regularly. These standards are even more important for structures such as schools and hospitals where the vulnerability of the occupants and the complexity and age of the buildings demand even more scrupulous attention to indoor air quality.

Most of all, the labor-management consultation on indoor air quality in the white-collar workplace must be just that: a consultation and a collaboration with everyone's rights respected. Unlike the industrial setting, white-collar environments usually involve portable work stations, which facilitates commonsense solutions to nonsmokers' complaints. In most offices, a formal smoking policy can be avoided by simple rearrangement of desks, offices, and work stations. Placing nonsmokers near fresh air ducts, providing portable smoke removal devices, and, as a last resort, providing separate smoking and nonsmoking areas will ameliorate any controversy over sidestream smoke exposure. All policies should be agreed to by all affected workers, and these policies should demonstrate fairness to all employee groups. It is important to remember also that policies that penalize clerical and junior staff for not having private offices are unfair and discriminatory.

If It's Broke, NEMI Can Fix It

Organized labor is taking other steps to prepare useful solutions to the indoor air quality dilemma. The National Energy Management Institute (NEMI) is a nonprofit corporation jointly managed by labor and management; its parent organizations are the Sheet Metal Workers' International Association (SMWIA) and the Sheet Metal and Air Conditioning Contractors National Association (SMACNA).

This industry has traditionally provided both the management and skilled labor needed to install and maintain HVAC systems. Following the OPEC oil embargos, the industry initiated several new programs to provide expertise in the energy management field to government and private building managers around the country. The 1981 establishment of NEMI was an outgrowth of this energy conservation purpose. Since then, the institute has established a National Air Quality Service subsidiary to facilitate the process of improving indoor air quality in the workplace.

As part of this activity, NEMI has developed training programs for contractors and technicians to acquaint them with the best methods of protecting indoor air quality. Through rigorous, in-depth technical studies, these personnel have become experts in every source of indoor air contamination and in the technical solutions best adapted to each indoor environment. NEMI markets this program, selects the contractor best suited to perform the work, provides project management oversight to ensure quality control, and serves as the client's contract manager. By offering complete building evaluation and repair services and by arranging financing for clients through shared-savings or lease-purchase agreements, NEMI has become the one source that is thoroughly qualified to help building owners and managers "cure" their sick buildings.

NEMI is also working with federal and state regulatory officials to develop guidelines and standards for safe indoor air quality and is in the process of developing a marketing program to acquaint potential clients with its services. This program is a landmark attempt by labor and management, working together, to promote the importance of indoor air quality and take concrete steps to provide a safer workplace.

Workplace Restrictions: A Slippery Slope?

Despite the development of NEMI and other innovative programs, labor would seem to be fighting a rear-guard action against increasingly restrictive smoking rules. Across the country, some private companies and many government agencies have already unilaterally instituted severe antismoking policies, including the banning of smoking on the job, refusal to hire smokers, and disciplinary action or even firing of workers who will not quit. But labor cannot refuse to protect the rights of its members who smoke; the war over smoking policy represents important terrain and some nonnegotiable principles for the entire labor movement.

For at least a generation, U.S. unions have recognized that their mandate to protect workers extends beyond the protection of wages, benefits, job security, and even health and safety. Protecting the workers also means protecting the workers' personal rights against arbitrary and unfair intrusion by management. This principle cannot be compromised in the debate over workplace smoking.

No one doubts that there are many antismoking activists, including some union members themselves, who sincerely believe the world would be a better place if everyone stopped smoking.

But there are also vegetarians who believe we would all be healthier giving up meat and exercise enthusiasts who insist that refusal to jog is tantamount to suicide.

The point is not whether any of these practices or their absence is personally harmful. It is whether these decisions properly belong in the jurisdiction of management in its role as employer. It is that claim to jurisdiction over the life-style choices of American workers that the labor movement must reject.

The right to personal choice is a characteristically American one, and its manifestations in the workplace are clear. Management has no right to dictate whether one should marry, who one's friends are, or what a worker's personal habits should be. One of the major purposes of the collective bargaining agreement is the protection of each worker's privacy and autonomy from the arbitrary whims and excesses of managers. The presumption of decision-making power must lie with employees, especially concerning personal life-style choices.

The dangerous precedent of arbitrarily instituted smoking regulations must be reversed. Some companies are going so far as to enforce hiring bans with polygraph tests, urianalysis, blood tests, or medical examinations—all major infringements upon individual rights. A few companies now prohibit smoking off the job—a nonsensical and threatening extension of an employer's control. When these controls are instituted as a result of nonsmokers' complaints, it is still the union's right and duty to negotiate a fair compromise that will protect all of its members. If the policy is imposed arbitrarily without worker complaints, unions must suspect a smoke screen that is hiding management's own dereliction of duty for providing a safe and clean indoor air environment.

The motivations for labor involvement in the entire debate over smoking and indoor air quality are complex but compelling. Protecting industrial workers from threats to their health and safety has been a mandate for unions for more than a century. Alerting white-collar employees to the new dangers in their supposedly clean environments is a modern challenge but still well within the context of union obligations to members. Taking a stand to protect the procedures of collective bargaining as the correct way to solve labor-management disagreements is also integral to unions' purpose.

But the most central, if sometimes difficult, duty of a union is to ensure that its membership is not exposed to arbitrary, random impositions and decisions of management. Traditionally this has meant that a company cannot lower wages, eliminate benefits, or fire workers without regard to the formal labor-management relationship. But in the 1980s, as the agendas for both labor and management become more economically and politically complex, it is crucial that the labor movement not lose ground in its fight for workers' dignity and autonomy. The resolution of the debate over workplace smoking will demonstrate organized labor's ability to fulfill this most important responsibility.

7

Dividing Is Not Conquering

A Manager's Perspective on Workplace Smoking

Jody Powell

As any manager knows, the biggest challenge in running a business is people—finding and keeping the right ones and ensuring their productivity and satisfaction on the job. As the chief executive officer of a growing Washington public affairs firm, I spend much of my time locating the paragons of virtue I need to run and staff our accounts. And while I'm searching for creative, experienced, detail-oriented, seventy-hour-a-week types, I don't want to worry about the personal habits of prospective and current employees that bear no relationship to the smooth functioning of my business.

To its credit, the American business community has learned to ignore differences that don't matter when dealing with personnel—class, race, gender, age, family situation, and what can only be called personal style. I'll hire an Albanian grandmother who rides to work on a Harley-Davidson if she can do the job. I am convinced that the evolution in corporate thinking toward policies that evaluate employees on only job-related factors brings greater fairness and productivity to the workplace. Managers who use their position to impose personal preferences on their subordinates ask for trouble. And the more intelligent and creative those subordinates are, the bigger the trouble will be.

Particularly ludicrous is the notion that I should ignore all extraneous considerations save one—whether or not my employees use tobacco products.

This chapter is an attempt to examine workplace smoking issues from legal and economic perspectives, as well as their impact on employee productivity and morale. I am neither a libertarian zealot nor a fanatic antismoker; I am a businessman trying to get the job done. I feel that the smoking issue has been blown far out of proportion by activists with little understanding and fewer facts on the controversy. I am convinced that workplace smoking can be handled by common sense, logic, and a good notion of a business's overall objectives.

The Legal Angles: Options and Pitfalls

Before making a decision about smoking policies in a business, it is important to understand what the law compels. The history of employee-initiated litigation on workplace smoking issues is complex, but for the most part it demonstrates a reasonable regard by the courts for employer latitude and common sense.

Although a meticulous interpretation of workplace smoking lawsuits should be left to the lawyers, here is a brief summary of the legal requirements placed on an employer:

Nonsmokers have no constitutional right to compel a smoke-free workplace. Although antismoking activists have searched for the right to a smoke-free environment as a concomitant of freedom of speech, due process, and the preservation of other unspecified "fundamental rights," federal courts have given short shrift to these arguments. In several cases in the 1970s, various district courts dismissed the notion of a right to smokeless air as nonsensical.

There is not much support in common law for attempts to compel a smoke-free working environment. One case, *Shimp versus New Jersey Bell,* granted a secretary the right to a smoke-free workstation; however, this case was never contested by the defendant, New Jersey Bell, in court. Attempts by antismoking activists to establish this case as a precedent have failed. The same state court that granted the *Shimp* ruling denied another complainant's request against the same employer several years later. Generally, courts have held that while employers are responsible for providing safe working environments, there is no correlation between "safe" and "smoke-free." Employees complaining of hypersensitivity to tobacco smoke are not guaranteed smokeless environments by reason of their own perceived needs or allergies.

Employees who smoke are entitled to consideration of their rights before smoking restrictions can be imposed. One court decision put this best by stating that "there is no warrant and no justification as a matter of civilized management to treat smokers as if they were moral lepers." For that matter, if there is a collective bargaining agreement with a group of employees, unilateral smoking restrictions *cannot* be imposed without risking litigation based on unfair labor practice. In other words, there may be as much potential legal liability from imposing a smokeless environment as in "neglecting" to provide one; a smart employer will look for compromise solutions.

That leaves the conscientious—or litigation-wary—manager with a feeling of some security from nuisance lawsuits. This does not mean that employers will not be sued by militant antismokers, only that demands for a smoke-free workplace based on federal or common law will probably not be upheld by the courts. There are, however, some exceptions where businesses can be held responsible for restricting workplace smoking, because of local ordinances or safety requirements specific to certain types of businesses.

The federal government, for example, has restricted smoking in government buildings since 1986. Most civil servants are permitted to smoke only in hallways and rest rooms; smokers who light up in their offices are liable to enforcement from the building's security officers. These restrictions are too new for any organized survey of the results to be available, but preliminary indications are that a great deal of tax-paid time is lost as a result of this prohibition against smoking and working at the same time.

Other businesses with a legalized antismoking policy include concerns where workers routinely handle food, flammable materials, chemicals, or sensitive equipment. In most of these situations, workers understand the smoking restrictions when they apply for employment and accept them as safe and necessary. No one objects to these restrictions in specified environments—just as no one objects to the wearing of protective clothing or the use of special procedures that are appropriate to certain types of businesses.

But in many states and localities, antismoking groups have initiated legislation that would deny freedom of choice to all employers and employees. At least eight states and fifty localities have passed statutes restricting or forbidding workplace smoking. As of 1987 there have been no cases of employees invoking such laws to demand smoke-free workplaces, but that possibility certainly exists in these jurisdictions. The results of litigation based on local ordinances is unpredictable and will depend to some extent on the stringency and ultimate enforcement responsibility written into each statute. But the possibilities do strike fear into the hearts of affected managers.

In the District of Columbia, for example, there is a statute on the books mandating no-smoking areas for restaurants and other public places. This law, despite the best efforts of antismoking activists, has yet to be extended to the private workplace. I consider it to be in the best interest of my own business to lend political support to those who would preserve my discretion over workplace policies and would urge other managers to take a similar stance in their own jurisdictions.

The Shell Game of Smoking Economics

For years, certain economists and efficiency experts have claimed that smoking on the job contributes to a loss of productivity and higher insurance costs. These experts support bans on workplace smoking in the name of cost-effectiveness; an argument that, if true, should concern every manager in every business or industrial environment.

But the true costs of smoking and/or smoking restrictions are difficult to discern. One advocate of restrictions, William Weis, serenaded the business community for years with claims that a flat refusal to hire smokers could "shave personnel costs by 20 percent, insurance premiums by 30 percent, maintenance charges by 50 percent, furniture replacement by 50 percent, and disability payments by 75 percent."[1] According to Mr. Weis, the total costs per smoker exceed $4,000 per year. That is a powerful bottom-line argument. But few executives acted on Weis's recommendations, and their skepticism appears to have been justified.

Weis admitted that "skeptics might argue that these numbers are as soft as the underside of a porcupine, and that may be true."[2] A close look at those numbers bears that out. Consider the statistics on worker absenteeism. Data from the 1976 National Health Survey demonstrate that male smokers are absent less frequently than nonsmokers; that work loss for women is lowest among women who smoke the most; and that women who smoke fewer than twenty-four cigarettes a day miss less work than former smokers. Moreover, statistics on absenteeism that measure only the smoking habits of employees fail to take into account other determining factors—age, gender, family responsibilities, personal problems, type of employment, job satisfaction, and even the weather for commuting employees. One obvious correlation missed by restriction advocates is that blue-collar workers smoke more than white-collar workers and have higher rates of absenteeism. Is this because blue-collar workers smoke or because they enjoy their jobs less and feel less involved in them than do their white-collar counterparts?

When considering increased insurance and medical costs, the correlation between smoking and higher premiums is shaky at best. Again, job category is the crucial factor. Blue-collar workers smoke more than white-collars; they also lift more, carry more, are exposed to dangerous industrial substances, and work with heavy and occasionally dangerous equipment. Premium rates for workers' compensation are determined not by employee smoking habits, but by occupational category, carrier experience with the business, and the statutory level for workers' compensation for the particular state. Marvin Kristein, an American Health Foundation economist who has claimed that the average smoker costs his or her employers between $336 and $601 a year, admits that "we lack meaningful 'case-controlled' company comparisons of experience with smoking employees versus nonsmoking employees versus ex-smokers and the impact on company costs."[3]

One issue that is not really quantifiable but about which exists informed opinion, is productivity. Despite allegations by antismoking groups that smokers accomplish less on the job than their nonsmoking counterparts, a survey of 1,900 supervisors has demonstrated that those who know their employees best do not concur. In that survey, conducted by Response Analysis Corporation, more than 90 percent of these first-line supervisors denied that employee smoking on work breaks affected their performance. Two-thirds of the supervisors denied any link between smoking *while working* and productivity; the one-third that did acknowledge a negative influence on performance are mostly in the manufacturing, transportation, communications, and utilities industries, where work is often manual.

Most of these supervisors were also negative toward smoking restrictions. Sixty-four percent had none in their businesses; those that did have formal policies cited safety, legal, and "aesthetic" (it might annoy the customers) reasons as motives for instituting the policy. But the overall attitude toward smoking employees was tolerant—only 6 percent of those queried agreed that a smoking ban in their workplace would allow them to accomplish the same amount of work with fewer employees. Only 3 percent believed that refusing to hire qualified applicants because they smoke was a sensible personnel policy.

When a manager evaluates the effect of restrictive smoking policies, it is not enough to deny a link between smoking and higher costs. Too little has been written about the negative effects on productivity, costs, and employee morale due to ***smoking restrictions,*** particularly in the white-collar environment.

Putting Up the Walls: A Costly Alternative

Antismoking policies come in all shapes and sizes. Some businesses limit smoking to designated areas. Others allow smoking only at designated times; some permit no smoking at all; and still others restrict smoking in the presence of clients or the public. Most (like my own) leave the issue up to the good manners and good sense of employees and negotiate any conflict between smokers and nonsmokers.

If a manager does institute some type of formal restriction, it is important to note the economic penalties involved. Telling employees that they may smoke only in the rest rooms, cafeteria, or hallways practically ensures that they will not be at their workstations as often as if they were permitted to smoke there. In an office environment, where there is no foreman or assembly line to compel attendance at a certain physical location, this means unanswered phones, untyped letters, unwritten proposals—undone work.

If smoking is limited by time, someone must devise a schedule of smoking breaks that staggers the workload done by professional and clerical staff. In my business, which usually demands ten-hour days, it would be impossible to create a schedule that recognized the clients' need for fast turnaround. Any business with time constraints will find that half its secretaries are on break when the final copy is due, or one-quarter of its account personnel is in the lunchroom when an emergency brainstorming meeting is called. I certainly don't want to send out bloodhounds for my supervisors when a late-breaking story affecting a client's interest comes over the wire.

Then there's the alternative that often works, but must be well loved by those in the building-remodeling business—physically separating smokers and nonsmokers. This policy is useful when flexibly implemented, but too rigid a set of restrictions can hamstring efficient operations. For example, every account group in my office has both smokers and nonsmokers. These people are in and out of each others' offices dozens of times a day; they share support staff, files, data-processing equipment, reference books, and sometimes, it seems, each others' brain stems. Like many managers paying high rent in urban locations, I have found that the only cost-effective and sensible way to arrange these groups is to have support staff share clerical bays, junior professionals share offices, and the whole account group share a corridor and meeting room.

Separating account groups by their smoking habits—at least on the basis of a formal, square-foot-specified policy—would lead to a nightmare of lost productivity. Obviously, if we have two junior account executives who smoke and work on the same accounts, putting them in the same office is a great solution. But smoking cannot be the determining factor in our physical layout, not without sacrificing the efficiency and quality of the work we do for our clients. This is not to say that we ignore nonsmokers' requests concerning sidestream smoke; we accommodate them as best we can. Investing in fans, portable air cleaners, and smokeless ashtrays often solves the problem of proximity between smokers and nonsmokers.

Any white-collar manager facing a space shortage, high rents, and the need to station employees by function has to reject rigid smoking restrictions for cost reasons alone. But the implications of antismoking policies go beyond the dictates of bean counting; the entire morale and esprit de corps of an enterprise can be compromised by directives that traduce employees' personal rights.

How to Offend Your Employees

Tell your employees what to do, what to think, and how to behave in areas that do not affect job performance. For example, one of my employees thinks it's barbaric to hunt ducks and turkeys—one of my favorite pursuits. In my view, she's not just soft-hearted but soft-headed on this issue, and I suppose that in hers I'm a practitioner of avian genocide. Does this affect her salary, promotions, job performance, or willingness to keep working here? Obviously, it does not.

I put the smoking habits of my employees—and for that matter, myself—in the same category. It is a traditional if unspoken American notion that the less we interfere with one another, the better off we all are. Rights in this country focus on the individual, not the government, the employer, or the opinions of other people. As I noted in the beginning of this chapter, a country whose legal and moral dictates do not permit employers to discriminate on the basis of age, gender, handicap, and so forth should think long and hard before legislating or dictating personal habits to adults.

Beyond the philosophical issues associated with any form of institutonalized smoking policy lies the practical problem of enforcement. If my business were located in a jurisdiction that demanded some type of smoking policy, presumably I would be responsible for calling in the cops if any employee continually broke the rules. Even a policy that was set up at my own discretion would require enforcement by me, another member of senior management, or the office manager. I shudder at the idea of disciplining adults—particularly talented, egotistical, creative adults—about a personal habit. Just trying to figure out an appropriate "punishment" for the "crime" of smoking demonstrates the ludicrous nature of the whole enterprise.

In the Response Analysis survey cited earlier, more than 63 percent of the supervisors queried believed that imposing any type of smoking regulations would worsen employee morale. Many sensible people concur but still worry about the charges hurled by antismoking activists from Surgeon General C. Everett Koop on down—that environmental tobacco smoke endangers the health of nonsmokers. But in looking carefully at the scientific data on sidestream smoke, I can only conclude that smoking is the least of a manager's problems in providing a safe and healthy indoor environment for workers.

Where There's Smoke, There's A Sick Building

Three recent scientific workshops on tobacco smoke in the air concluded that the evidence on negative health effects on nonsmokers is, at best, inconclusive. Without delving into the scientific jargon, it's best to quote one of these reports: "Should lawmakers wish to take legislative measures with regard to passive smoking, they will, for the present, not be able to base their efforts on a demonstrated health hazard from passive smoking."

This conclusion, from an international workshop held in cooperation with the World Health Organization in Vienna in 1984, sums up the current state-of-the-art evaluation of the effects of tobacco smoke in general. But an even more important source of specific data is available. Two researchers from the Harvard School of Public Health, who examined various indoor environments for the presence of carbon dioxide and nicotine, discovered insufficient amounts of either to affect the health of nonsmokers. The amounts of carbon dioxide ranged from one-tenth to one-fifth of the levels permitted by the Occupational Safety and Health Administration; the amounts of nicotine were so small that a nonsmoker would have to spend 100 straight hours in the smokiest bar to inhale the equivalent of one filtered cigarette.

This doesn't mean that those of us working in clean, nice, white-collar environments are free of workplace hazards. On the contrary, the newer and more energy-efficient our building is, the more likely it is that we are breathing unclean air. And that threat would exist even if every smoking colleague were summarily dismissed.

When the National Institute of Occupational Safety and Health examined more than 200 buildings in response to occupant complaints, they found that fewer than 2 percent of these complaints could be traced to tobacco smoke. The complaints were *not* unfounded—poor indoor air was indeed causing headaches, chronic respiratory infections, sore throats, and other symptoms. The most common cause was inadequate circulation of fresh air. Other causes of poor indoor air quality included the circulation of indoor contaminants, such as the fumes from copying machines, and the circulation of outdoor contaminants, such as carbon dioxide from underground parking garages.

What does this mean for the manager? It means that our responsibility to provide safe workplaces for employees goes well beyond a faddish fascination with smoking restrictions. Most office buildings erected since the energy crises of the 1970s are built for fuel efficiency. It is in the economic interest of our landlords and building-maintenance firms to ventilate these buildings with clean air as infrequently as possible. Since the windows can't be opened, and the ventilation systems are controlled by a computer, we as tenants often have no apparent means to get more clean air into our offices. In the building in which we rent space, for example, the building engineer opens the vents to allow outdoor air to circulate only once a month, which is once a month more than permitted by the building management firm that employs him.

Indoor air quality problems can be serious. There have been actual epidemics of respiratory illnesses in "tight" facilities. Legionnaire's Disease and certain types of pneumonia are directly attributable to bacteria-infested ventilation systems. A manager who ignores the symptoms of "sick building syndrome" in space he rents or owns can be liable for the effects on the health of employees.

Responsible managers, if faced with concerns about unclean air, should monitor office air quality, if necessary, by calling in private air-quality experts or local government inspectors. With firm data to prove that the building is inadequately ventilated, we can then negotiate with the building owner or maintenance firm to improve the quality of the air and replace ineffective heating, ventilation, and air-conditioning equipment.

This responsibility cannot be avoided and should not, because solutions of air-circulation problems will often solve any workplace smoking controversies as well. If tobacco smoke is lingering in an indoor environment, it is probably a symptom of a much bigger problem. Unclean, poorly circulated air is a health problem whether or not anyone smokes. Smoking restrictions will not solve it. But ensuring clean, well-circulated air will almost always solve the question of smoking restrictions to everyone's satisfaction.

To Convince a Zealot

Even if a manager threads his or her way through the legal, economic, psychological, and scientific pitfalls of the workplace smoking issue, a few crusaders will always thwart the most reasonable attempts to work out solutions. I want to respond to an anecdote recounted by a gentleman whose ability to be dogged and even a bit abrasive on issues far thornier than smoking is well documented. He is also, I should add, a gentleman who has become a friend in the years since my White House tour.

In Sam Donaldson's book, *Hold On, Mr. President,* he proudly relates the tale of his crusade to ban smoking in the White House press room. Fortunately, this particular confrontation occurred during the tenure of my successor as press secretary. As Sam tells it, he waged a fearless battle against Larry Speakes and other reactionary defenders of polluted air in an effort to get smoking banned from the entire press area.

I know very well that the White House press room is a rabbit warren with little room to breathe or move around under any circumstances. I sympathize with nonsmokers who work there for hours at a time. But I also sympathize with smokers who are chained to their desks under the same circumstances. To chase them out in the cold while they're waiting for a story to break does not seem to be a fair solution to the problem. Satisfying the militants at the expense of everyone else is not a solution that Sam would advocate for the Middle East or the deficit crisis.

Whether it's the White House press room or a more spacious suite of offices, the ethics of the situation remain the same. By enlarging space, putting up room dividers, investing in air cleaning equipment, or working to improve the air circulation in the entire building, managers have the technological tools with which to quench the workplace smoking controversy.

Most important, managers have as allies the common sense and courtesy of their staffs, who should be encouraged to negotiate about smoking issues the way they must about other shared concerns. Thirty percent of adult Americans smoke. Restricting their rights is not the path to harmonious working conditions. Dictating standards of personal behavior that do not affect job performance to employers and employees alike has far-reaching consequences, none of which are in the interest of American business or American workers.

Notes

1. *Personnel Administrator*, May 1981.
2. Ibid.
3. *Preventive Medicine*, March 1983.

8

Environmental Tobacco Smoke and the Press

R. Emmett Tyrrell, Jr.

On December 16, 1986, the U.S. surgeon general released a report on environmental tobacco smoke. The way its findings were mangled shows the press at its most pliant. Upon publication of the report, the surgeon general heaved in his obiter dictum that the report proves environmental tobacco smoke is harmful to nonsmokers. That statement is untrue, yet the press dug no further. One wonders if reporters even read the report. Now I am a nonsmoker. I may smoke an occasional cigar while writing a piece—not this one!—but for the most part I remain smokeless, and other people's smoke at times annoys me even more than other people's cheap cologne, trivial conversation, and rude manners, the last of which, incidentally, can cause fisticuffs and worse. As it turns out, the surgeon general's report does not provide irrefutable evidence of environmental tobacco smoke's mortal peril. In fact on no fewer than 147 occasions, according to my reading, the report demurs from making this claim and actually points out that it cannot conclusively establish such a claim.

Nonetheless, in the United States, the press coverage of the surgeon general's findings reported the story as though the report had amassed unassailable proof of the mortal danger of environmental tobacco smoke to innocent, smokeless bystanders. The *Boston Globe* asseverated, in obvious error, what the rest of the press implied in its December 17 edition: "Breathing other people's cigarette smoke causes uncounted cases of lung cancer and other diseases among nonsmokers, U.S. Surgeon General C. Everett

Koop declared yesterday in a report to Congress." Elsewhere in the world wherever the press reported the story, the same inaccuracies were reiterated, though the story was much less of a headline outside the United States. Concern about public smoking seems to evanesce the farther one travels from the United States, but that does not mean that the misreported surgeon general's report will never be an issue outside the country. The move to render cigarette smoking first controversial, then dangerous, and finally illegal, or at least regulated, began in the United States. It has grown in strength, however, throughout the rest of the world. The press has usually been immensely helpful in spreading this concern.

To understand the press's mindless and apparently unscotchable dissemination of exaggerated claims against environmental tobacco smoke, one must understand a very important element in the culture from which these rumors have spread. The United States, and in varying degrees all other nations sharing its liberal, progressive values, has a culture that is always militant on behalf of one perceived noble cause or another. The United States is by nature reformist. Its citizens are never resigned to any discomfort, material or otherwise. They tinker with government, with society, with health. No human activity passes unnoticed by American reformers. American society is forever marshaled to do things: to end poverty, to end criminality, to spread literacy, goodwill, health. You name it, whatever at one time or another has seemed wholesome, there have been movements of Americans ready and willing to change things for the better. This is true of other parts of the world too. Since the eighteenth century, the reformist spirit has been a catalyst for change in Western civilization, and its effects are felt to one degree or another wherever the vectors of Western civilization have spread.

The press is one of the leading institutions of reform in the United States and throughout the rest of the West. It is mission oriented. Its first mission is to discover and disseminate the news. Why? What if the news is irrelevant to an audience? What if it is embarrassing or disturbing? It is sent out anyway. And what if it is inaccurate? That is, what if a fact is wrong? Perhaps the facts are right but the emphasis is wrong. Many journalists are not daunted by any of these questions. They know that they serve a high-minded purpose: general enlightenment and improvement.

Whatever is seen at the time as improving and enlightening humanity is thus considered news, and this news must be reported in such a way as to fetch the interest of the audience. It also must move the audience to improving society. Thus we see that the very concept of news is reformist.

Those who dig it up, prepare it for an audience, and send it to the audience, either over the airwaves or in print, have an affinity for reform. Reformist issues catch the fancy of the press and get reported as reformers would have them reported. All that is necessary is that a position gain general acceptance as being reformist.

The prohibition of smoking is now such a position in the United States, and it is becoming a reformist position elsewhere. The press can be counted to be on the side of the opponents of smoking. The press can be counted on to read a surgeon general's report contradicting his charge that environmental smoking causes cancer and report the finding in the way reformers would like that finding to be reported. The press will overlook all the report's qualifications and assert that the report argues that environmental tobacco does cause mortal illnesses.

The press's position on smoking is highly irresponsible. Its position is a threat to civil liberties, for in the United States smokers are being oppressed on the basis of flimsy evidence. What is more, an important institution is discrediting itself. That the press has overstated the threat of environmental tobacco smoke as presented in the surgeon general's report is only one of many examples of the press's allowing itself to be used by reformers.

One might object that the press represents diversity and that there are many different points of view represented in the press. In the United States, there are not, most likely for the reason stated above. The press's predilection for seeing itself as a reformist institution and for being by its nature somewhat reformist has conduced it to adhere pretty much to a moderately left point of view. This domination of the press has been clearly demonstrated by scholarly studies, the work of S. Robert Lichter and his colleagues being most renowned.

In their superb study of America's prestige press, *The Media Elite: America's New Powerbrokers*, they show an amazing disparity between the prestige press and the mainstream. For instance, in the four presidential elections nearest the survey, the journalists voted overwhelmingly for the Democratic candidates. Lyndon Johnson received 94 percent of the journalists' votes in 1964. Hubert Humphrey received 87 percent in 1968; George McGovern, 81 percent in 1972; and Jimmy Carter, 81 percent in 1976. Nearly half the journalists believe the government should guarantee jobs. Fifty-six percent believe the United States exploits the third world. Fifty-seven percent believe the United States is immoral in its use of resources.

Thus we should not be surprised by the bias of the press once it gleans the party line on an issue. The party line once established is repeated endlessly. The party line on environmental tobacco smoke is that it kills thousands of people every year. Never does the press state that most studies on the connection between environmental tobacco smoke and cancer have failed to establish even a statistical relationship, much less a causal connection. Moreover, a nonsmoker would have to sit behind a desk or in a restaurant for hundreds of hours nonstop to be exposed to the vaporous equivalent of one cigarette. The credulous press accounts of cigarette smoke's carcinogenic effects on nonsmokers put one in mind of nothing so much as the press's exaggerated reports of the carcinogenic effects of such substances as saccharine, which have to be introduced into laboratory animals in vast quantities before the animals develop tumors.

The selectivity of the press toward reporting on tobacco smoke can be brazen. My favorite example of this sort of selectivity comes from the November 20, 1986, issue of the redoubtable *New England Journal* in which an article asserted that the danger from smoking was not reduced by a reduction in cigarette smoking. That article fit well into the reformers' anticigarette campaign. The issue also contained an article describing studies that showed "a reduction of as much as 50 percent in the relative risk of endometrial cancer among female smokers, confirming previous evidence of a protective effect of cigarette smoking on this disorder." This article did not fit nicely into the reformers' anticigarette smoking campaign, and the press generally ignored it.[1]

The press worldwide tends to settle on an interpretation of an issue, an event, or a person and repeat it endlessly. This tendency has also affected the press's bias against the cigarette. I first encountered the tendency far from the issue of cigarettes and health. While researching a book in the late 1970s on American public figures (*Public Nuisances*), I discovered that with each figure, certain anecdotes one put into print would be forever repeated. This was true of personality traits. No matter the vicissitudes a figure might pass through, no matter how many later anecdotes might pile up to entertain and illuminate readers, the press would always tend to repeat the earlier material.

The most amusing and perhaps flagrant example of this tendency to repeat a party line is the case of the late Major Claude Robert Eatherly, who became a public figure in the United States after a credulous newspaperman came across him languishing in a Texas jail. He was a winner of the Distinguished Flying Cross. His career in the U.S. Air Force was one of constant achievement, and after World War II he was headed to the top. But then he felt a twinge of conscience about that mission over Hiroshima, and the twinge became a profound guilt complex. Soon he became self-destructive, and his behavior landed him in the Texas jail.

Once the story was out, Eatherly became a celebrity, invited to reveal his psychic wounds before antiwar rallies and on television. Bertrand Russell expounded on Eatherly's moral significance. He was awarded the 1962 Hiroshima Award "for outstanding contributions to world peace."

Then the truth came out: the writer William Bradford Huie discovered that Eatherly never won the Distinguished Flying Cross, had not commanded the Hiroshima bombing mission, and had not bombed the Japanese even with conventional bombs. He went through all of World War II without having fired a hostile shot. His Hiroshima experience amounted to flying a navigation plane over the fated city to collect weather samples. Later he competed furiously for the honor of dropping atomic bombs in the first Bikini tests. Unfortunately, he did not even qualify for that brush with the atomic age, and soon he was bouncing around the United States in repeated scrapes with the law.

That is the historic record on Claude Robert Eatherly. Yet after his July 1, 1978 death, the *New York Times* reported, "Claude Robert Eatherly, who, as a young Army Air Corps pilot, picked a hole through the clouds over Japan in his bomber on the morning of August 6, 1945 and radioed the B-29 Enola Gay to drop its atomic bomb on Hiroshima, died of cancer last Saturday in Houston. Even in death his legend grew. According to the *Times*, one of the dead fraud's brothers said that Eatherly had probably been exposed to radiation as a government pilot during the Bikini atoll nuclear tests.

And so should any reader be surprised to hear that the press reports on the 1986 surgeon general's report misstated the report's conclusions on environmental tobacco smoke? The press's misstatements on this matter will most likely be repeated until the end of time or until some other good cause seizes the reformist fervors of the fourth estate. The process is not particularly rational. As the campaign against smoking is now going, cigarette smokers are at best going to be sequestered in special areas in restaurants and workplaces. At worst, all indoor smoking will be banned. Yet the campaign is irrational. The danger of smoke from indoor smokers is insignificant. And while we are on the subject of sequestering smokers, consider this: there are two high-risk groups that carry AIDS, a deadly infectious disease that is incurable. Thus far the reformist press has shown no interest in discriminating in any way toward those high-risk groups prone to carry the disease. The press campaign against environmental tobacco smoke is capricious, intolerant, and ignorant. But it will continue.

Note

1. Payton Jacobs III et al. "Influence of smoking fewer cigarettes on exposure to tar, nicotine, and carcinogens," *New England Journal of Medicine* 315 (20 November 1986): 1310–13.

9

Environmental Tobacco Smoke

Ideological Issue and Cultural Syndrome

Peter L. Berger

As a result of a systematic and, on the whole, successful campaign by the antismoking movement and its bureaucratic allies, the American public has become increasingly conscious of the issue of environmental tobacco smoke (ETS, sometimes also called *passive smoking*). The success of the campaign is to be measured not only in terms of public awareness but in a panoply of government actions (local, state, and federal) seeking to limit ETS in public places and in the workplace. One additional measure of success has been the increasing aggressiveness of antismokers, who have felt progressively freer to harass, and occasionally even assault, smokers in restaurants, waiting rooms, offices, and other places where ETS is alledgedly a problem.

Seen in the perspective of the antismoking campaign, a perspective frequently reflected sympathetically in the media, the issue is quite simple: not only is ETS an annoyance to nonsmokers (an "invasion of their private space"), but it is, more important, a matter of public health. ETS, albeit in a less intensive way, promulgates the same diseases that smoking supposedly does. Smokers thus constitute not only an annoyance but a health hazard to nonsmokers. The scientific basis for this belief is alledgedly conclusive, having been legitimated by an authority no less than that of the U.S. surgeon general. Anyone who questions the belief, therefore, is either motivated by wishful thinking (smokers, in the main) or by vested interest (the tobacco industry). A health hazard having been established, government action is obviously called for. The campaign, on all levels of government, seeks progressively to limit and eventually to eliminate ETS in all places where smokers and nonsmokers mingle. The openly avowed end result is to proscribe smoking in any public area.

A sociological look at the issue discloses a much more complicated reality. The approach of the sociologist must rigorously bracket the question as to the scientific validity of the alleged evidence because he or she can claim no expertise in physiology or epidemiology. An important point to make, though, is that very few people indeed can claim such expertise, yet many people have passionate convictions in this matter. These convictions, in the nature of the case, cannot be based on scientific reasoning; rather, as with most other beliefs, they are based on faith in authorities. But since authorities conflict, the indidiviual chooses which authority to give credence to. Thus it is at least noteworthy that people who believe very little of what government agencies tell them on other important matters (foreign policy, for instance) will regard an issue as closed because the surgeon general has spoken on it. The sociologist will want to know why faith in the surgeon general is plausible to people who will not believe one word uttered by the secretary of state, and this question may be asked quite apart from the scientific question of who is right in the final analysis, an analysis that remains inaccessible to the vast majority of people who are neither medical nor foreign policy experts.

Only the most unimaginative (and, one may add, uninformed) rationalist will assume that most people's beliefs are based on the results of science. If that assumption were correct, human history since, say, Newton would have taken a very different course. One may well regret this reality, but one will not understand any human society unless one reckons with it. But to say that most beliefs are based on faith is not to relegate them to a realm of mystery inaccessible to further exploration. People's beliefs are rooted in a culture, which can be described and understood. And very often beliefs are directly linked to vested interests, in which case sociologists have used the term *ideology* to designate them. An ideology is a system of beliefs that serves to justify the vested interests of its adherents. In what follows ETS will be discussed as an ideological issue in this sense and also as a reflection of much broader cultural trends.

In the many years of controversy over smoking and health, the phrase *vested interests* has been used almost exclusively to refer to the tobacco industry, and it is again so used in discussions of ETS. No objective observer is likely to quarrel with this appellation. One may well imagine that the tobacco industry would be very pleased if there were evidence to show that ETS helps cure depression, the common cold, or some other condition that ails people. To concede this obvious point, however, in no way implies that there are no vested interests, no ideology, on the other side of this debate. (Only very rarely does one come upon an issue in which no vested interests are entangled; typically, this would be an issue of utmost obscurity in which hardly any one could possibly be interested.) The vested interests in this case are very clear, and they are powerful.

There is, most obviously, the vested interest of the antismoking movement itself. This is no longer a little band of lonely zealots. Rather, the movement is large, well organized, and providing employment as well as status to sizable numbers of people. The movement has been successful in enlisting powerful allies in miscellaneous bureaucracies both public and private, such as those of federal agencies and of voluntary organizations such as the American Cancer Society. Many individuals in these bureaucracies have developed strong vested interests in this issue, even if those interests are less total and less zealous than the interests of movement people. Anyone who attended the several world congresses on smoking and health staged periodically by the World Health Organization of the United Nations could see in action what by now constitutes an international antismoking conglomerate. One would have to be extraordinarily naive not to perceive that these people have a vested interest in this issue. Indeed it is not fanciful to suggest that, one by one, every individual with an interest in the proposition that ETS is harmless could be matched by an individual with an interest in the opposite proposition. Finally, both movements and bureaucracies have a specific location in the larger society, one typically characterized as one of class. This too pertains here.

What is the ideological function of the ETS issue? The freely proclaimed agenda of the antismoking campaign is to stigmatize, segregate, and (at least partially) criminalize smoking. An often announced goal of the campaign is to reduce smoking to a socially unacceptable activity engaged in by consenting adults in private. Smoking, in other words, would then be an activity comparable to what at least in the past was seen as repulsive sexual practices, morally condemned by almost everyone, driven into secluded private spaces, and punished by law if it dared to emerge from these spaces—ideally, in the imagination of antismoking activists, "the habit that dare not speak its name." (One cannot help wondering whether human communities do not require some such odious activity to feel morally righteous about, so that if one is given up, another must take its place.

The same people who quiver with indignation at the pejorative designation, let alone proscription, of even demonstrably unhealthy sexual practices will happily push for the criminalization of smoking. Thus California appears to be in the forefront of resistance to any AIDS-prevention measures that might infringe on the right of gays and in the forefront of measures to outlaw smoking. People who believe in the rationality of history have a lot to worry about.)

This antismoking agenda has run into a considerable problem from the beginning, at least in Western democracies with a strong tradition of individual rights (and, in any case, thus far the antismoking campaign has had much less success in other parts of the world). Smoking, after all, is a voluntary activity; no one is forced to smoke. Thus, even if smoking really causes diseases, it could be argued that this is a risk freely shouldered by smokers, who have the right to live their own lives as they choose and who must be solely responsible for any unfortunate consequences to themsleves. In the earlier history of the antismoking movement, when its main agenda was to alert ("educate") smokers to these alleged health hazards, this was not much of a problem. But as the movement increasingly sought government action to proscribe and regulate smoking, a solution to this problem had to be found. One can say a priori what shape such a solution had to take: some way had to be found to argue that smokers constitute a danger not only to themselves but to others as well. If so, of course, a doctrine of individual rights could no longer stand in the way of coercive government action.

Perhaps inevitably, then, a new figure was introduced into the debate, that of the innocent bystander. Thus, the ETS issue introduces a whole new category of victims into the debate over smoking. Not surprisingly, this issue has occupied an increasingly important place in the antismoking propaganda, and antismoking activists have been particularly angered by any public statements of scientists or others to the effect that the evidence on the dangers of ETS is less than overwhelming.

There have been other victimological themes in the antismoking propaganda, with specific sets of putative victims: children, both born and unborn (and children are "innocent" by definition); women (feminists have made this point within the antismoking movement, though they have been caught in an ideological double-bind on this; the data indicate that women smoke more as they enter the labor force, and entry is favored by feminists); blacks (who in the United States proportionally smoke more than whites); and the third world (where smoking rates have been climbing consistently and where the tobacco industry, in line with conventional "third worldist" rhetoric, can be portrayed as but another instance of capitalist exploitation). But ETS has one great victimological advantage: here, supposedly, is a threat not to any specific set of people but to everyone; thus everyone becomes a victim. The smoking threat is effectively universalized. (To coin a phrase, one might refer to this as "victimological maximalization.")

One further convenient aspect of the ETS issue is that the antismoking propaganda habitually refers to smoking as an "epidemic"—in its more exuberant rhetoric as "the largest preventable epidemic in the world." Needless to say, there are competitors in the field—not only for public attention but for funds, public and private, and for governmental priorities, all matters of great interest to the antismoking conglomerate. Put simply, the ETS issue helps keep smoking close to the top of the "epidemics" list in the relevant circles of discourse (say, in the ambience of the World Health Organization).

Where there are ideological gains, there are also ideological losses. In the campaign, there have been two approaches to smokers: one aggressive and punitive and the other benevolent and therapeutic. In the former, the smoker is treated as an enemy, in the latter as a victim-patient. Over the years, the emphasis in the antismoking propaganda has tended to shift from the former to the latter approach, because, not surprisingly, it has proved more effective. Many smokers are embarrassed about their habit, would like to reduce or quit, feel guilty and vulnerable. An avuncular approach here is obviously more productive than a prosecutorial one. The ETS issue, though, tends toward aggression and punitiveness toward smokers. In the context of this issue, virtually inevitably, the smoker is cast in the role of villain. This invites counteraggression. It is too early to say how far this will go, but there are already scattered newspaper accounts of smokers going on the offensive to protect what they consider to be their rights.

Finally (though this can only be touched upon here), there are wider class interests involved in this issue. In Western countries, there are strong class differences in smoking behavior. The working class smokes more than the middle class. What is more, the antismoking attitude is most strongly represented among the most educated segments of the upper middle class (segments sometimes designated as the New Class or the knowledge class—broadly speaking, the intelligentsia). This stratum has a collective interest in government as against the private sector because, compared with other segments of the middle class, it derives more of its income and status from government expenditures and programs. But, be this as it may, there is also the matter of class culture. Sociologists have accumulated a vast amount of data showing the differences in beliefs and behavior between different classes, ranging from religious convictions to sexual practices. It comes as no particular surprise, then, that classes should differ in smoking behavior. However, one important characteristic of the upper middle class in general and its intellectual segment in particular has been its propensity to inflict its own values and behavior patterns on the lower classes—by "missionary" work but also, if possible, by governmental coercion. Prohibition is the best-known example of this in U.S. history. The antismoking campaign is yet another instance of this coercively inclined bourgeois imperialism. In a society with a democratic regime and a strong antielitist tradition, there are probably built-in limits on such an enterprise. Again, it is too early to tell how much of a problem this class struggle aspect will constitute for the antismoking campaign.

There are, of course, movements and political processes that operate in relative isolation from the larger society and its culture. They are rare. In this instance, it is plausible to see the antismoking phenomenon as part and parcel of much wider sociocultural developments. For what remains of these observations, then, the focus will shift from the ETS issue and its ideological ramifications to the antismoking phenomenon as a cultural syndrome—that is, as a constellation of attitudes and actions that disclose much more general features of contemporary Western (and especially American) society. This can be done by looking at several themes that point from the antismoking issues to the wider society.

Foremost are the themes of health and youth. The antismoking phenomenon in a very obvious way is linked to a pervasive concern for health and for the preservation of youth. It is important not to exaggerate the contemporaneousness of this theme. After all, it is safe to assume that human beings in any society or period of history have preferred to be healthy rather than sick and have wanted to prolong the vigors of youth. In the contemporary West, however, and especially in the United States, the concern for health and youth has become a collective obsession. Indeed, it is possible to speak of a cult of health and youth. For a considerable number of people, it appears, remaining healthy and youthful have become central values, which take precedence over just about any other considerations.

Advertising on American television is perhaps the clearest, most dramatic expression of this. A huge portion of it is devoted to the propagation of products and services that promise to keep those who purchase them healthy and youthful. The precise focus of these concerns is subject to fashion (and presumably to shifting commercial interests). In consequence, there are shifting fads in this area. The jogging mania is one example; another is the rush to health food. There is, of course, a tangible dark side to this: the fear of disease and disability that is to be kept at bay by these allegedly healthy-promoting activities. There is, after all, little pleasure for most people in running sweatily and breathlessly through urban traffic or in subsisting on diets from which any remotely agreeable food has been carefully removed. Rather, one jogs and diets in order to prevent heart disease. At least until the advent of AIDS (which, so far, no one has causally linked to smoking), cancer was the most fearful disease of all. Consequently an antismoking life-style is propagated as a defense against that disease, which is feared by most people. This fact helps to explain the ferocity of the antismoking campaign and the depth of the latter's antagonism to the tobacco industry ("purveyors of death") and, to some extent, against ordinary smokers. Thus the antismoking activist who harasses a smoker in a restaurant or some other public place is implicitly (sometimes literally) saying: "Stop killing me!"

An important aspect of this concern for health and youth is moral. Health and youth are not only desirable; they are also moral imperatives. This has the simple but far-reaching implication that the individual who is not healthy and youthful has only himself or herself to blame—unless, that is, blame can somehow be assigned to this or that set of malevolent others. Both possibilities are very much in play in the antismoking phenomenon. The evidence indicates that many smokers feel guilty. They are not only doing something that, they believe, might be harmful to them (that, by itself, would make them anxious but not guilty) but something that constitutes a moral weakness. This psychological conjunction of fear and guilt makes smokers very vulnerable indeed. It helps to explain the surprising wimpishness with which they have responded to the aggression directed against them by the antismoking campaign. If, on the other hand, one does not want to blame oneself, one must blame others. The tobacco industry has served as the obvious target for the assignment of blame, allowing the anxious smoker to be appealed to as victim rather than be pilloried as villain.

Another important and close-to-the-surface theme is environmentalism. Again one must be careful not to exaggerate. In a complex technological society, concern for the environment is a perfectly rational matter, especially because many environmental problems (such as air pollution) are directly experienced by ordinary people in their daily lives. The environmental movement, however, has produced rhetoric that goes far beyond such practical concern. It has evoked a utopian vision of a society that will benefit from modern technology without having to pay the costs—a vision of a harmonious, healthful world in which William Blake's "dark Satanic mills" have been demolished once and for all. This theme of bucolic utopianism has found many expressions and causes since it achieved public prominence in the late 1960s: the antinuclear movement, the campaign against the automobile, the propaganda against fluoridation (which began earlier) and asbestos, and others. The antismoking campaign fits snugly into this universe of ecological causes. It has used a good deal of ecological language: "smokers' pollution," "smoke-free environment," "personal space," and the like. The ETS issue lends itself to this linkage with particular ease, as already indicated by its name.

Less salient but still significant is the theme of consumer protection. Tobacco evidently comes in products purchased by consumers, and the protection of the latter against putative negligence or worse by the producers is a theme carried by the antismoking propaganda. In the United States, with its exuberantly generous product liability laws, this has been an important legal aspect of the antismoking campaign, with individuals suing cigarette manufacturers for damages allegedly incurred to their health as a result of smoking. (Thus far, it appears, the courts have been much less amenable than legislatures and executive bodies to the arguments of antismoking groups.) Quite apart from the legal aspect of this theme, it also contains an anticapitalist element. Tobacco is big business. Tobacco products are manufactured by large corporations, many of them multinationals. Casting the tobacco industry in the role of malevolent, death-purveying power comes most naturally to people on the political left. There is insufficient evidence to assess how important this anticapitalist element is within the antismoking movement in Western countries. Other information about the ideological and political inclinations within the class that serves as the social context of the antismoking movement would tend toward a hypothesis that anticapitalist sentiments will reinforce antismoking beliefs and activities, but such a hypothesis remains to be tested. The anticapitalist-antimultinational animus has come out most forcefully in those parts of the antismoking propaganda dealing with the third world and its alleged exploitation by the tobacco industry.

There are strong emotional themes in this syndrome. To be an antismoker is not just an intellectual position leading logically to certain activities;

it involves strong, sometimes passionate feelings. These have already been touched upon in the observations made about the moralization of health. Relevant emotions are fear, anxiety, aggression, and a yearning for security. Given the belief that smoking causes disease and early death, there is an obvious basis for fear and for aggression against those who allegedly promote or deny this danger. However, one may hypothesize that smoking serves as a socially available focus for free-floating anxieties and, in consequence, scapegoating aggression that may have different or quite diffuse origins. All of this pretty much adds up to what Richard Hofstadter has called the "paranoid style in American politics." There is a pervasive paranoid texture to the antismoking phenomenon, and this emotional trait reminds one of many other movements in recent (and not so recent) U.S. history. Once again, prohibition and the temperance movement preceding it come immediately to mind. This too could be phrased as a hypothesis, yet to be tested: smoking serves as a focus for a variety of free-floating anxieties and aggressive impulses. Put differently, many people in modern society are frustrated and angry without being quite sure what causes this; they are now offered a causal explanation of what ails them and a legitimate target for aggression.

Smokers constitute a minority, and a shrinking one—apparently around 30 percent of the adult American population. This may be, to some, a plausible minority to vent aggression upon. It is too large a group, though, to be safely hostile to; it is more economical if the anger can be concentrated on the tobacco industry. Still, once one thinks of smokers as a minority, it is instructive to reflect how one minority after another has been removed from the American scene as a legitimate object of hostility. There is, then, a certain emotional logic to the scapegoating of smokers. Thus one may observe how individuals who would be outraged by discrimination against racial or religious minorities as well as minorities of life-style preferences other than smoking appear to have no moral qualms in advocating violent discrimination against smokers—including physical segregation (literally "in the back of the bus"), denial of employment, denial of insurance, and sundry modes of public humiliation.

A submerged but probably very significant theme is that of risk avoidance—a salient theme in contemporary society, as Aaron Wildavsky and Mary Douglas have pointed out. Compared to any other society or period of history, the contemporary West is a safe place to live—no matter whether one measures safety in terms of life expectancy, hunger, disease, crime, or government oppression. Despite this (and perhaps because of this) contemporary Westerners (with Americans in the lead) are extraordinarily fearful. It seems that every few months the media offer up yet another terrible danger, hitherto hidden from view, that is about to engulf us. Some of these dangers are undoubtedly real; others are the product of paranoid imagination. But a psychologically plausible reaction to this array of putative dangers is the quest for a risk-free existence. This expresses itself in a great variety of areas, from the avoidance of parenthood (children are arguably the biggest risk of all) to the avoidance of military means in international relations. There are, of course, different ways in which individuals can protect themselves against risks. They can live life cautiously (the so-called survivalists are an extreme case of this, but it is useful to see them as an exaggerated version of a much more general attitude). They can band together with others, in a "movement," to stave off the danger. They can appeal to the law (the litigational propensity of Americans is at least partly explained by this). And, last but not least, they can turn to the government for protection. The expectation that the welfare state ("Papa State," as the Germans put it) will protect the individual against every conceivable risk is a potent political reality in Western democracies. All these risk-avoiding methods are much in evidence in the antismoking syndrome.

It is perfectly rational to avoid unnecessary risks, and human beings have always done so. What comes into view here, however, is a utopia: the notion of a risk-free life. It is a utopia because it is beyond possible attainment. Put differently, what we see around us is not only an unwillingness to take risks but a steady lowering of the risk threshold—that is, of the notion of what constitutes a reasonable risk. To paraphrase Hofstadter, one might call this the "paranoid life-style in American culture." The obverse side of this generalized paranoia (or, if one prefers, hypochondria) is precisely the utopia of the risk-free life: a vision of unending health and youth—indeed, in its final philosophical intention, a dream of immortality. Once again one may hypothesize here: a secular redemption serves here as a substitute for religion in a group of people (the college-educated upper middle class of Western societies) in which traditional religion has lost much ground. Secularized Western culture has generated its own peculiar version of the ancient quest for the secret of eternal life. Gilgamesh appears among us in a white coat, telling us with all the authority of science and government how we may live forever. His admonition to give up smoking is but one, relatively small, part of his overall message.

The antismoking phenomenon in general, and the ETS issue in particular, must thus be seen in a wider cultural and political context. It bears repeating that none of the foregoing observations prejudge the scientific question of whether smoking or ETS is or is not hazardous to health. But the social context within which this question is being debated must be taken into consideration if one is to understand what is going on. Antismoking beliefs and activities are part of a much broader cultural syndrome; they do not exist in isolation. They also involve vested interests, which generate an appropriate ideology. And these vested interests, of both antismoking activitists and bureaucrats, relate to wide social conflicts, most of them rooted in the dynamics of class. The antismoking phenomenon, especially in the United States, constitutes one battlefield in a much larger cultural war, or *Kulturkamph*. As so often in history, this *Kulturkamph* can also be understood as a manifestation of *Klassenkampf*, or class conflict. Here, as in many other areas of contemporary social life, the upper middle class, and especially its cultural elite, is arrayed against other classes in a struggle for power and privilege. It is sociologically plausible to predict that the outcome of the debate over smoking will hinge, at least in part, on the outcome of this larger sociocultural conflict.

10

Prohibitions and Third-Party Costs
A Suggested Analysis

Burt Neuborne

THERE is a rhythm in the evolution of a prohibition. Whether the taboo involves forbidden sex or forbidden food or forbidden pleasure, would-be controllers of behavior generally begin by urging potential actors to refrain from the target activity because it is morally or culturally wrong. When the prohibition is another culture's, we are quick to recognize that the definitions of right and wrong underlying the prohibition are themselves often cultural artifacts that reflect the needs of powerful elites in a society far more accurately than any objective set of moral criteria. Not surprisingly, when the prohibition is our own, we are less likely to recognize its equivocal source.

Once a prohibition has become entwined in the moral texture of a culture, its enforcers will almost always seek to demonstrate that engaging in the disfavored behavior is physically, as well as morally, harmful to the actor. At this stage in the evolution of a prohibition, science is conscripted in aid of morality in an attempt to increase the persuasive force of the taboo. Armed with scientific evidence that the target behavior is physically harmful, enforcers of the taboo seek to persuade actors that they should refrain from the activity for their own physical, as well as moral, good. Given the initial sense of moral or social unease about the behavior, scientific evidence proving its physical danger often finds an enthusiastic, and uncritical, response. Witness the extraordinary consensus in the scientific community during the late nineteenth century about the harmful effects of masturbation.

Once the scientific evidence reinforcing a taboo reaches a sufficient level of public acceptance, behavior controllers intensify their efforts on two levels. First, they seek to ensure that potential actors are fully aware of the scientific evidence of the consequences of their folly. In a free society, one cannot quarrel with such an effort. One must, however, be cautious about a general tendency to overstate the certainty of the scientific evidence linking the target behavior to physical harm and a companion tendency to attack anyone who questions the evidence. When science and morality march hand in hand, it is difficult to challenge either partner.

Second, as the scientific evidence reinforcing the prohibition becomes more universally accepted, behavior controllers seek to eliminate inducements to engage in the target behavior by forbidding speech that portrays the behavior in a favorable light. If potential actors can be bombarded with scientific evidence that the behavior is harmful and, at the same time, be insulated from any encouragement to engage in the activity, the prohibition may be enforced without the necessity of direct controls on behavior. Such a self-enforcing prohibition is the controller's dream since it shields from public debate the question of whether there should be a prohibition by retaining the illusion that actors are voluntarily refraining from the target activity.

Whether such indirect prohibition through information control is successful depends, first, on the degree of consensus about the danger of the activity to the actor and, second, on the strength of a society's commitment to the values of personal autonomy and free speech. When a society's resistance to indirect prohibition through information control is too great, generally because values of personal autonomy take precedence in a particular culture over scientific evidence of harm to the actor, behavior controllers turn to direct legal prohibitions on the target activity. But precisely because they have been thwarted in an attempt to achieve indirect prohibition because of the society's commitment to free speech and personal autonomy, advocates of direct prohibition cannot safely rely upon the harm the activity may cause to the actor as a basis for an enforceable prohibition. Saving individuals from hurting themselves by engaging in foolish behavior is a notoriously weak basis for imposing direct controls on a disfavored activity in a society that places a high premium on values of personal autonomy. Rather, would-be behavior controllers must up the ante by arguing that the activity not only harms the actors but endangers innocent third parties as well. When reformers sought to require motorcyclists to wear safety helmets, their most persuasive argument was not the stupidity of risking one's life by riding without a helmet but the effect on third persons of preventable accidents. Significantly, in our cultural and legal history, neither sermons about the moral degeneracy of drinking nor warnings about the harm it causes the drinker were sufficient to trigger the legal prohibition of alcohol. Rather, it was widespread acceptance of the assertion that alcohol harmed nondrinkers that pushed prohibition over the top.

Unless behavior controllers can transfer attention from the harm that a disfavored behavior causes the actor to the harm it allegedly causes third persons, the chances of imposing direct legal prohibition on the behavior are quite slim in any culture that places a high value on personal autonomy. Recognition of that fact puts a tremendous premium on the identification of third-party costs in any campaign to prohibit a controversial activity. Since effective prohibition turns on the third-party cost issue, the same scientific community that was mined for evidence that the disfavored practice is harmful to the actor will be asked to opine on whether the practice is harmful to third persons. The obvious danger is that the scientific consensus on harm to the actor will infect scientific judgment on the issue of possible third-party cost. After all, political science describes more than a college major.

If a controversial practice is perceived as morally questionable as well as harmful to people who engage in it and if the only way to achieve an effective prohibition of the practice is to demonstrate third-party costs, two predictable phenomena will emerge. First, the scientific community will be under substantial pressure to validate claims of third-party costs; and, second, advocates of prohibition will seek to use the third-party cost issue as the basis for direct prohibition. The process is rendered even more volatile when the disfavored activity is engaged in disproportionately by racial or sexual minorities. It is much easier to tell relatively weak segments of the society that they must stop disfavored behavior because it allegedly threatens everyone else than to tell the powerful that they are hurting the weak.

Assertions about the third-party costs of controversial behavior should therefore be viewed skeptically when they are advanced as the last phase in the evolution of a prohibition—but not because they cannot be true. No one can deny that some controversial behavior hurts not only the actor but innocent third parties. Legal prohibition of such activity does not raise troublesome philosophical issues. But precisely because the terminal phase of a prohibition occurs with years of momentum and reformist fervor behind it, particular care must be paid to the scientific foundation underlying the claims of third-party harm, especially where the target behavior is one that is engaged in disproportionately by weak actors in the society.

Moreover, merely because a third party's sensibilities are offended by certain behavior does not justify suppressing the behavior. We have refused to permit hurt feelings or aesthetic disapproval to count as a third-party harm in the prohibition equation—and rightly so. Third-party preferences based on aesthetics or sensibilities should not qualify as a basis for prohibition. Allowing aesthetics or sensibilities to count as a real third-party cost justifying prohibition would vastly enlarge the scale of potentially banned behavior while introducing a subjective and culturally determined factor as the last step in the prohibition process. Thus, allowing unsubstantiated claims of physical harm to third persons to serve as the basis for a legally enforceable prohibition would open the door to casual prohibition based on class-bound aesthetics.

Smoking, as a form of controversial behavior, illustrates virtually every phase in the evolution of a prohibition. From its inception, smoking was viewed with suspicion by moralists and arbiters of taste, who saw it as both hedonistic and aesthetically unacceptable. Moral concerns about smoking graduated into very real fears about the health risks of tobacco. Once a critical mass of scientific consensus was reached on the health issue, it became difficult to challenge the scientific data without being subjected to sanction. Predictably once the orthodoxy of the scientific condemnation of smoking was generally accepted, powerful attempts to ban speech portraying tobacco in a favorable light were introduced in Congress. Once that effort was stymied by the culture's deep attachment to free speech and individual autonomy, the attempt to base direct legal prohibition on third-party cost rapidly took center stage. Campaigns to ban smoking in a host of public settings to protect nonsmokers from the danger of so-called passive smoking became the order of the day.

I lack the scientific expertise to evaluate claims that the alleged harm to third persons caused by smoking in public places justifies legally enforceable bans on public smoking. I do feel confident, though, in placing the third-party cost claims about smoking in the broader context of the evolution of prohibitions generally. Viewing the debate over the alleged third-party costs of smoking in that broader context yields several observations.

First, care must be taken to evaluate claims of third-party cost with a skeptical eye in order to be certain that the claims are anything more than old arguments for prohibition presented in a new wrapper. In the absence of a genuine scientific consensus about the third-party costs of public smoking, prohibition may well be justified to save smokers from themselves, but it cannot fairly be justified as necessary to save third persons from harm. While a scientific consensus on the third-party cost of public smoking may evolve, the present state of scientific opinion appears to be decidedly mixed on the issue. Thus, if the significant line between prohibitions designed to save actors from themselves and bans aimed at protecting third persons is to be preserved, bans on smoking cannot be justified as necessary to protect third persons, at least under the current state of scientific knowledge.

Second, the aesthetic preferences and sensibilities of third persons should not count as a real third-party cost in assessing the justification for a legally enforceable prohibition. Since the initial impetus for many prohibitions is a dislike of the activity by a powerful segment of the society, allowing aesthetic preference to count as a third-party cost would transmute prohibitions that are really designed to save people from themselves into restrictions that are glibly justified as necessary to protect others. The net effect of such a shift in thought patterns would be a quantum increase in the ability of the group to dictate the behavior of the individual. The danger of allowing aesthetics to qualify as a third-party cost is particularly acute when the proposed prohibition is aimed at an activity disproportionately engaged in by racial or sexual minorities.

Third, although it does not appear that the scientific case can be made at the present time for prohibitions on public smoking designed to protect the health of nonsmokers, the impact of public smoking on nonsmokers should not be ignored. Nonsmokers are genuinely troubled by fears that passive cigarette smoke may be harmful to their health. Moreover, while aesthetics and sensibilities should not count as a third-party cost in weighing the need for a legally enforceable prohibition, the fact that smoking in public places is annoying to large numbers of people is a fact that must be confronted.

It is possible, of course, to treat the issue of smoking in public as an all-or-nothing proposition. Either the preferences of nonsmokers must trump the desires of smokers or vice-versa. That is the way all too many disputes about prohibitions are resolved. I believe that such all-or-nothing resolutions are appropriate only when it is impossible to construct solutions to a problem that provide tolerable recognition of the interests of third persons while respecting the preferences of persons who wish to engage in the controversial activity. When smoking in public is the issue, it should almost always be possible to craft compromises that will shield third persons from undue exposure to unwanted behavior while permitting smokers to engage in desired behavior. The operative goal should be to maximize the behavior preferences of all participants rather than to sacrifice the preferences of one group to the desires of another.

Finally, I do not intend this essay as an argument against banning tobacco. Although I am skeptical about prohibitions generally, I accept the proposition that some behavior is so harmful to the actor that society has the right to forbid it. Whether smoking tobacco falls into that very narrow category of behavior raises difficult questions about whether society should attempt to protect individuals against the perceived consequences of their self-destructive behavior. I would probably vote against a ban on tobacco because I do not believe that society should be in the business of protecting people from themselves except in the most extreme settings. I am certain, though, that my vote would turn on whether I thought that society should dictate to individuals how they should live. I would not take the easy way out and impose a prohibition in the name of protecting third persons against alleged dangers that have not yet been documented in a sufficiently persuasive manner.

11

Smoke

A Politician's Angle

Lord Bruce-Gardyne

EMERSON's judgment on the political animal—"the louder he talked of his honour, the faster we counted our spoons"—was unfairly severe. Politicians in modern democracies are motivated, like other wage slaves, which for the most part they are, by two considerations: job security and the hope of preferment. Theirs is an exceptionally insecure environment and differs from that of other wage slaves in that their jobs are more at risk from the performance of others than from their own behavior and that preferment is given and withdrawn according to criteria that in any other walk of life would be deemed irrelevant or even absurd. They secure office because they go over well on television and lose it because they are caught in sexual infidelities, which their fellow citizens indulge in with impunity. Their lives—and the lives of their families—are conducted in the glare of publicity; that is their privilege and also—not infrequently—their misfortune. They must, by definition, commit themselves to the service of their fellow citizens, and that commitment is more often genuine than not. It does not mean (although occasionally it may) that they are out to swipe our spoons. It does mean that their first and overriding concern is to get themselves and their party reelected.

These may seem like blinding glimpses of the obvious. Yet it is curious, in my experience as a practicing politician of over twenty years in the United Kingdom and a government minister in the U.K. Treasury Department, how often they are overlooked or misunderstood by those who seek to influence the formulation of public policy. They may be cynical about the real motivations of their politicians. Yet paradoxically they are prone to take too easily at face value the philosophical stance of those politicians, a stance that, in practice, will almost always take second place to their calculations of reelectibility and preferment.

In the 1980s, respect for the judgments of the marketplace has enjoyed a remarkable revival across the globe. After more than thirty years of a fashion for state corporatism, the Thatcher government in Britain and the Reagan administration in the United States led the way in cashing in on a reaction against the degeneration of corporatism into hyperinflation and the demand for consumer choice. The example they set, and the evidence of its electoral effectiveness, encouraged imitation in surprising places: in the third world and even in the Communist bloc.

Logic might suggest that governments elected on a ticket of respect for consumer choice would refrain from legislation, fiscal manipulation, and arm twisting designed to limit or direct that choice. Logic, however, is a bad guide to political behavior. In numerous areas governments elected on market-oriented platforms have proved no less assiduous in ordering the lives of their fellow citizens than the corporatists who preceded them (indeed in certain respects they have gone further). Legislation for the compulsory wearing of seat belts in cars has become commonplace; family choice with regard to abortion is increasingly restricted; most obviously the discouragement of smoking by all means short of outright prohibition has continued to gather momentum.

Why should these things be? Essentially, I suggest, because politicians in a modern democracy respond to two—not infrequently conflicting—stimuli. Reelection depends on success in satisfying the aspirations of the mass electorate, no doubt. But while reelection is a necessary precondition for preferment (even in some presidential systems, such as that in France, where government has to command a parliamentary majority), it is not a sufficient condition. Another important condition—there are others that need not concern us in this context—is a politician's standing with the communications media. This is the channel by which nonconsumerist, and sometimes anticonsumerist, pressures can be brought to bear on market economies.

We live in a health-obsessed environment. In most advanced industrial countries, the fading of religious belief has led to the elevation of longevity to the rank of an unquestioned public good. Our forebears accepted the teaching of the church that while life might be "nasty, brutish and short," it was the hereafter that really mattered anyway. Today we have no such reassurance, and the church itself is as keen as anyone else that we should defer the hereafter as long as possible. Indeed the pursuit of longevity sometimes edges close to the pursuit of immortality. I remember once listening with mild incredulity to one of my colleagues in the House of Commons warning us that "beyond a doubt heavy smokers will die." I interrupted to ask him what would happen to the rest of us. It was regarded as an intervention in the worst possible taste.

Similarly, when I was serving as a U.K. treasury minister and had to listen to another colleague calling for penal increases in tobacco taxation on the grounds that smoking-related ailments imposed immense and unnecessary costs on our national tax-financed health service, I departed from the text of the response prepared for me by officials to point out that an infinitely heavier burden was laid upon our health service by our modern propensity to live beyond the biblical span, often into years of senility that might be miserable and were certainly enormously costly to the taxpayer. Even my officials thought I had gone way over the top. Yet the latest estimates suggest that in the United Kingdom, septuagenarians now cost the health service some $1,500 apiece each year to provide for; and as we survive into our eighties and nineties, we are into serious money—perhaps three times that figure. Nor should it be overlooked that pension entitlements are still generally related to actuarial calculations of life expectancy terminating in the mid-seventies; and hence that longevity imposes an uncovenanted burden on the pension system too.

Such calculations do not discourage politicians with an eye to preferment from identifying "healthy living" as an idea whose time has come—notwithstanding the fact that their electors regularly disregard it in their daily lives. There is nothing new to this dichotomy. Denunciations of adultery have featured in political rhetoric down the ages however much the electors—and the denunciators themselves—indulged in it. I recall one member of the British Parliament who devoted his weekends to participation in temperance marches in his electoral district and had to be assisted into taxis nightly during the week at Parliament by the discreet attendants. What is different about some of the modern "health" issues—and that of smoking notably among them—is that their protagonists are also motivated by hostility to the profit motive.

It is striking, but by no means coincidental, that France is the one advanced industrial country where smoking increases and the real incidence of tobacco taxation declines. In France the domestic production and marketing of tobacco is a state monopoly. It is a highly profitable monopoly; but in France the state looks after its own. Besides, many of those who work in the communications industry, and in particular in television, reserve their hostility for private enterprise profits. The profits of state enterprise tend to be emotionally sanitized. It is true that some crusaders go to the lengths of accusing national governments of living off immoral earnings because of their revenues from the taxation of tobacco. But the obvious contradiction between such accusations and the clamor—from the same people—for the aggressive use of taxation as a deterrent to smoking inevitably blunts their impact. The "poisonous profits" of private enterprise tobacco manufacture and distribution represent a less ambiguous target.

This combination of aggressive propaganda from the health lobby and the instinctive hostility of many of those in the communications industries toward corporate profitability has created a social climate in which smokers have become an oppressed minority. Almost like the homosexual community of old, they are made to feel ashamed of their propensity. They do not ask for favors, and therefore the politicians know better than to be identified as the defenders of their interests and consumer preferences (notwithstanding the fact that those interests and preferences, unlike those of the homosexuals of old, are universally legitimate).

Hence those of us in politics who object to state intervention in matters of legitimate personal choice are, in this instance, obliged to look elsewhere for supportive arguments better attuned to the selective intolerances of our age. Fortunately these are not lacking.

Top of the list, obviously, is the economic significance of the tobacco industry itself. The "health" lobby knows that politicians from the state of Virginia, or a country like Botswana where tobacco production is a major source of local wealth and employment, are beyond their reach. But in many of the major consumption markets, production of the raw material is either marginal or nonexistent, and manufacture is concentrated in a small number of locations, and even there employment in the industry—and hence its influence on voting propensities—is declining. In these markets influence upon the attitudes of politicians needs to be carefully targeted.

Some years ago when I was serving in the U.K. Treasury, the British tobacco industry organized through its retail outlets (the thousands of mostly independent High Street shops selling newspapers, cigarettes, and confectionery) a national petition in the run-up to our annual budget-making process in the spring, designed to bring home to politicians in general, and through them to treasury ministers in particular, the hostility of voters toward discriminatory increases in tobacco taxation. I am afraid it was a waste of time and money.

It was a waste of time and money because politicians are profoundly cynical about petitions. They confront them all the time, and they know that signatures upon them do not remotely imply commitment on a scale to sway voting intentions. It is always easier to sign than to decline to do so.

This, however, does not mean that producer interests are powerless to sway political attitudes. But they have to generate more direct pressures at what might be called the political points of sale. The farming industry has long offered the example of how this can be done. Throughout the Northern Hemisphere, the farming vote is hardly more than marginal, and shrinking, and yet, as we all know, it achieves a net transfer of resources from the urban majority to finance surplus food production. This is essentially achieved because farmers regularly take time out to wait upon their local legislators in impressive numbers. The local legislators emerge powerfully motivated by the instinct of self-preservation to champion farmers' interests on Capitol Hill or at Westminster. Other producer lobbies, including the tobacco industry, have been rather slow to learn from this example. In the Westminster Parliament, some members representing traditional tobacco manufacturing towns such as Nottingham and Bristol do speak up in defense of the industry in Parliament. Many do not. But if the thousands of small retailers, to whom cigarette sales are an important element in the viability of their businesses, could be induced to lobby their local legislators as the farmers do, the case for the industry might not have to depend on a few lone voices from the manufacturing centers.

An ancillary producer argument that ought to be persuasive is one that, in my experience, has hitherto been surprisingly underexploited. This is the vital importance of the tobacco crop to some of the poorest third world countries. The "health" lobby has made considerable headway in enlisting the support of church and charity interests against the "immorality" of smoking; yet these are also interests that are preoccupied with the plight of third world countries. The devil should not be allowed to get away with all the best tunes.

We must never forget, however, that while in the democracies the elected politicians are constitutionally supposed to call those tunes, it is usually the bureaucracies that write the music. Skillful lobbying may persuade the politicians to change the scores prepared by their officials for them to sing to, but the surer route is often to seek to influence the way the score is written in the first place.

Since taxation is the point of greatest direct governmental impact on the fortunes of the industry, the revenue department is the obvious first priority. To revenue departments, if my experience in helping to run one is any guide, tobacco is a milk cow, neither more nor less. Revenue departments are not much swayed either by the arguments of consumer choice, on the one hand, or by health or environmental considerations, on the other. They simply want to milk the cow.

All potential milk cows, in the budget-making season, convert themselves into geese that lay golden eggs, in danger of extermination. Revenue departments have a ready answer for all of them: tables to demonstrate that while sales may have shrunk as a result of their depredations, revenues have grown. They reckon they can calculate to a nicety the point at which the law of diminishing returns would set in.

But tobacco is special. In the eighteenth century, the British philosopher-politician Edmund Burke observed that "to tax and to please, no more than to love and to be wise, is not given to men." Revenue departments identify an exception to this serviceable rule of thumb. Taxing tobacco pleases many and is even perceived by many of the users as a fit punishment for their misdemeanor. The revenue departments find the temptation irresistible.

Apart from the awareness that they must not push their luck too far, their hands can be stayed by one other consideration: the effect of an increase in the cost of smoking on the national inflation index. The reason why the tax on cigarettes was not adjusted, even for inflation, in the 1987 British budget, was quite straightforwardly the impact of nonadjustment upon the cost of living index in the run-up to a parliamentary election. No wonder the "health" lobby campaigns in many countries for the exclusion of cigarettes from the calculation of the index.

But the tobacco industry, like any other under siege, needs a friend at court. Just as a president, vice-president, or secretary to the treasury from the state of Virginia must be advantageous in the United States, so a powerful secretary of state for Northern Ireland in the British cabinet is an inestimable ally, for in Northern Ireland viable employers are not thick on the ground; tobacco is one of them. This means that not only is the Northern Ireland bureaucracy a vigilant defender of the industry; the U.K. Treasury is also influenced, for it is always under pressure to commit resources to the subsidization of enterprises in Northern Ireland from both employment and counterterrorist considerations. For the most part, such enterprises are perceived by the Treasury as being doomed to failure.

All the more reason, in its eyes, not to inflict unnecessary damage on one industry that does not call for subsidy. But the effectiveness of this consideration depends crucially on the clout that the Northern Ireland secretary is reckoned to carry in cabinet.

The other bureaucracy with which the tobacco industry has to deal—that responsible for public health, which regulates advertising and sponsorship—is pretty much of a forlorn cause. In the present environment, officials who support the concept of consumer choice (and they do not breed abundantly in health departments anyway) keep their heads down. Exposure of such libertarian inclinations would land them swiftly in the public pillory. Venality is the best counterweight. Many of what are esteemed as socially desirable activities, from football to the live theater, depend increasingly on sponsorship for survival. But they are not robust in their defense of their right to make their choice of sponsors and are not to be counted on. The industry must look elsewhere.

This is an anthology of contributions to the debate about environmental tobacco smoke, and I have got thus far without mentioning the subject. This is because ETS is, so far as I can judge as one of them, not an issue that has seriously impinged upon public policy attitudes among British politicians. To date. It may come. But it is, in any case, no more than an extention of the controversy we have had to live with in the public policy arena for almost thirty years: the controversy between freedom of legitimate personal choice and the desire to impose restraints upon that choice in pursuit of longevity. I have sought to describe the nature of those countervailing pressures on the politicians as I have seen them and the nature and motivation of their responses to them. Like Mark Antony, I do not seek to praise or blame. Like Martin Luther, "I can no other."

12

Politics and Meddlesome Preferences

James M. Buchanan

EACH of us has a preferred pattern of behavior for others, whether they be members of our family, our neighbors, our professional peers, or our fellow citizens. I prefer that my neighbors control their children's noise making and disposal of their tricycles; I prefer that these neighbors refrain from rock music altogether, and that, if such "music" is to be played, the decibel level be kept low. I prefer that their backyard parties be arranged when I am out of town. I also prefer that my neighbors plant and maintain shrubs that flower in May for my own as well as their enjoyment.

I do not, however, exert much effort to enforce my own preferences on my neighbors' behavior. I trust largely to their own sense of fair play, common decency, and mutual respect. I do this because I know that my neighbors, also, have their own preferences about my behavior. They prefer that I control the barking of my dogs and that, if dogs must bark, that this be allowed only in normal hours. The neighbors also prefer that I refrain from operating my chain saw or power mower early on Sunday mornings.

There is an implicit recognition by all parties here that, although each may have preferences over the others' behavior, any attempt to *impose* one person's preferences on the behavior of another must be predicted

---

Reprinted by permission of the publisher, from *Smoking and Society* edited by Robert D. Tollison (Lexington, Mass.: Lexington Books, D.C. Heath and Company, Copyright 1986, D.C. Heath and Company).

to set off reciprocal attempts to have one's own behavior constrained in a like fashion. An attitude of "live and let live," or mutual tolerance and mutual respect, may be better for all of us, despite the occasional deviance from ordinary standards of common decency.

Such an attitude would seem to be that of anyone who claimed to hold to democratic and individualistic values, in which each person's preferences are held to count equally with those of others. By contrast, the genuine elitist, who somehow thinks that his or her own preferences are "superior to," "better than" or "more correct" than those of others, will, of course, try to control the behavior of everyone else, while holding fast to his or her own liberty to do as he or she pleases.

Private Spaces

I commenced this chapter with reference to my own personal relationships with my neighbors. Each person could fill in his or her own bill of particulars that would be descriptive of his or her own set of "social interdependencies." The general point to be emphasized is that such "social interdependencies" are necessary elements of life in civil society and that such interdependencies cannot be eliminated even in the most idealized allocation and assignment of "individual rights" among separate persons. A somewhat different way of putting this point is to say that there are no self-evident "natural boundaries" that define the "private spaces" within which individuals may be allowed to behave as they wish without affecting the utility or satisfaction of other persons, whether negatively or positively. Robinson Crusoe on the island before Friday arrives is useful as an expository device precisely because such a setting of total social isolation can *never* be experienced by an individual in a society.

To say that social interdependencies must always be present is not to argue that the assignment of property rights to persons cannot be of great value in reducing the potential for conflict among persons. As Thomas Hobbes recognized, the definition of what is "mine and thine," combined with the coercive power of the state to enforce the boundary lines so drawn, allows individuals to get on with producing their own goods rather than fighting over goods that belong to no one. The point is that no matter how carefully drawn and detailed is the assignment of rights, there must remain some potential for conflict. The fact that my preferences extend to your behavior over activities that are well within your defined rights, and vice-versa, ensures that my satisfaction is influenced by the way that you behave and that your satisfaction is also affected by my behavior.

The extent, range, scope, intensity, and importance of the social interdependencies among persons depend on the characteristics of the setting. The hermit in the forest may approach the Crusoe extreme. The frontiersman who ventures to the trading post only once a year remains largely independent, behaviorally and socially. By contrast, the suburbanite who lives in the townhouse must affect and be affected by the behavior of neighbors, fellow commuters, fellow consumers in the shops, those from whom he or she purchases goods and services, fellow workers, and numerous other separate interacting groups. So long as his or her allowable activities are well defined and enforced, the suburbanite coexists with others in the urbanized society without undue potential for overt conflict, retaining his or her preferences over the behavior of other persons in many of the roles that he or she confronts. But the suburbanite also proceeds to behave within his or her own well-defined and legally protected sphere of behavior on the presupposition that this sphere will be respected by others, that the set of rights he or she possesses will not be subject to invasion, either by other persons or by the collectivity as a unit. Civil order is described by each person "doing his own thing" within the limits of his assigned "private space," even when each person recognizes that some elements in his or her behavior will affect the satisfaction of other persons and that, reciprocally, the behavior of others, again within their own "private spaces," will influence his or her own well-being.

The Emergence of Potential Conflict

A potential for conflict may emerge from any one of several sources. The social juxtaposition of persons from totally divergent cultures may destroy the behavioral reciprocity that normally characterizes stable civil order. (For example, the alleged behavior of Asian political refugees in eating dogs created major social tensions in long-established communities.) The explicit violation of established patterns of behavior for the sheer purpose of attracting attention may create antagonisms that were not previously recognized. (For example, the flaunting of manners, hairstyle, and dress in the 1960s, primarily by the young, opened up the generational conflict that remained present in the 1980s.) An increase in moral fervor accompanied by conversion to life-styles that are dictated by a "new religion" may make tolerance for contrasting life-styles more difficult to accept.

My concern here is not with these, or other, possible sources of potential conflict among persons, families, and communities in the areas of social interdependencies. From the fact that we do have preferences about the behavior patterns of others, there is always a latent potential for conflict. And, as with other preferences, the attempted expression of these will depend on the relative prices required for their satisfaction. I shall suggest later that the relative prices of satisfying our preferences over the behavior of others are dramatically reduced, in an apparent sense, by the overt *politicization* of social interdependency. Before exploring this process more fully, it will be useful to examine nonpolitical means of adjusting to potential conflicts in areas of social interdependence.

Conflict Resolution through Voluntary Adjustment

Because each of us has preferences over the behavior of others in many separate social interdependencies, there remains always a potential for conflict. There is no guarantee that tolerable levels of mutual adjustment in behavior will be acceptable to all parties in an interaction. It may well be necessary to initiate or to undertake actions aimed at a voluntary resolution of the potential conflict. I may feel intensely negative toward the life-style of my neighbor, a life-style that does not allow me to invoke the laws of nuisance. If my preferences about his or her behavior patterns are so important to me as to suggest initiation of action on my part, there are several avenues open. I may make some effort to bribe or compensate my neighbor to modify his or her behavior in ways more pleasing to me. (To economists, this would be the avenue suggested by the Coase theorem. To noneconomists, this would perhaps seem one of the least plausible approaches to the problem.[1]) Or I may take actions that will reduce the spillover harms that the behavior exerts on me: I may install sound barriers, for example, to keep out the sound of rock music. As an ultimate step, I may consider shifting my location to a new set of neighbors, and, indeed, this potential for residential, locational, occupational, professional, purchasing, selling mobility is one of the most attractive features of the American society by comparison with those of other more rigid structures. The mere existence of effective alternatives, even if I never choose to exercise the exit option, ensures that there are relevant thresholds of spillover effects that cannot readily be crossed. These thresholds are important for each of us, and their existence surely helps make life in society tolerable.

The point to be emphasized about each and every one of these voluntary adjustments to potential conflict among interacting persons and groups is that the satisfaction of preferences over others' behavior within their legal rights is *costly*. That is to say, if my neighbor does not act in accordance with my preferences, I can either compensate him or her, build protection against the damage, or move. Each of these activities involves costs to me, and this cost will ensure that my interest in my neighbor's behavior is important enough to make the outlay worthwhile. There is a great difference between being merely irritated at the behavior patterns of my slovenly neighbor and actually paying him or her to "clean up his act." My "meddlesome preferences," to use Amartya Sen's expression, can be satisfied only at a positive opportunity cost.[2]

Conflict Resolution through Politics

There is no such rough matching of costs and benefits when the resolution of conflicts in social interdependence is approached through political mechanisms. If my neighbor's behavior irritates me, but not sufficiently to make it worthwhile to seek voluntary resolution, I may still be quite pleased if the town council will pass a regulation outlawing the behavior in question. It will cost me little or nothing to vote for the prospective councilman who will promise to outlaw leaf burning on the lawn; I may gain perhaps a few cents worth of utility on one or two autumn afternoons by imposing, through politics, my preferences on the actions of my neighbor. The costly steps that might be required in the absence of political institutions seem to be avoided. It seems that I can impose my own preferences on others at relatively low prices.

The cost saving here is only apparent, however. If I can resort to politics to impose my own preferences on the behavior of others, even if these preferences are not highly valued intrinsically, then it would seem that other persons, in working democratic process, can do the same to me. I may find that the political process is double-edged. If it can be used to my advantage in imposing my personal preferences over the behavior of other persons, it can be used to my disadvantage in imposing the preferences of others on my own behavior. I may gain a few pennies worth of utility by the regulation against leaf burning but find that possessing a handgun in my house is politically prohibited. And it may happen that I very strongly value the liberty to possess a handgun. The political process, which is allegedly open equally to all citizens, is evenhanded here.

It generates a few pennies worth of utility to me in restricting my neighbor's leaf burning; it generates a few pennies worth of utility to my neighbor by outlawing the possession of handguns. But, in so doing, it imposes many dollars worth of loss on me through preventing my possession of a handgun and imposes many dollars' worth of damage on my neighbor who highly values the liberty of burning his or her own autumn leaves.[3]

The Partitioning of Political Issues

The central thrust of my argument should be clear. The majoritarian institutions of modern democratic politics are exceedingly dangerous weapons to call upon in any attempts to reduce conflicts in areas of social interdependence. They are dangerous precisely because the institutions are democratic and open to all citizens on equal terms: what is sauce for the goose is sauce for the gander. Unless the person who calls upon politics can ensure that he or she retains some monopoly of political power, his or her own preferences are as likely to be imposed upon as imposed.

This danger inherent in democratic institutions tends to be overlooked because political decisions are partitioned so that each potential conflict is handled separately and one at a time. The interdependencies among the separated political decisions tend to be obscured and overlooked, with the result that it is quite possible that *all* persons will be placed in positions less desired than those that would be present in the total absence of politicization.

This central point may be illustrated by concrete examples, all of which have been at least partially politicized at one level of government or another in recent years in the United States or in other Western countries. Consider the following politically orchestrated regulations:

1. Prohibition on private leaf burning.
2. Prohibition of the possession of handguns.
3. Prohibition of the sale or use of alcoholic beverages.
4. Prohibition of smoking in public places or places of business.
5. Prohibition on driving or riding in an automobile without fastening seat belts.
6. Prohibition on driving or riding on a motorcycle without wearing crash helmets.

The listing here could be expanded greatly if we should add activities that some persons or groups have advanced as candidates for politicization. The six activities listed are, however, sufficient for my purposes here, and these are all familiar examples.

It seems quite possible that, at least in some political jurisdictions, a majority of voters might be found to support each and every one of the six activities listed. As noted earlier, however, the critical weakness in ordinary majoritarian procedures is that the intensities of preference are not taken into account. A bare majority of voters may support the prohibition on handgun possession, but a small minority may value highly the liberty to own handguns. The same result may hold for each of these activities. Yet, because the issues can be isolated and considered one at a time for political action, all of the regulations listed might secure majority support. But the handgun owners may find their loss of liberty much more valuable than the very mild feelings of benefits they secure from having the other activities prohibited. The same thing may hold for those who value intensely the liberty of drinking or smoking. The political process may well work so as to make each and every person in the relevant community worse off with enactment and enforcement of all of the prohibitions listed than he or she would be if none of the prohibitions was enacted.

There is a message in my argument here. Let those who would use the political process to impose their preferences on the behavior of others be wary of the threat to their own liberties, as described in the possible components of their own behavior that may also be subjected to control and regulation. The apparent costlessness of restricting the liberties of others through politics is deceptive. The liberties of some cannot readily be restricted without limiting the liberties of all.

The "Scientistic" Mentality

Critics of my argument here can charge that I have discussed the dangers of using the political process to impose one set of private and personal preferences over the behavior of others as if these preferences were mere whims, analogous to my dislike of the hairstyles of the youth of the 1960s. These critics might suggest, with respect to the set of activities listed above, and others, that the sumptuary prohibitions or regulations need not reflect purely private preferences. These prohibitions and regulations, existing or proposed, may be based on "scientific grounds." These critics might allege that leaf burning releases dangerous elements in the atmosphere; that handguns kill people; that alcohol is addictive and a causal factor in disease; that smoking is dangerous to health; and that seat belts and crash helmets save lives.

These arguments are highly deceiving in that they attempt to introduce, under the varying guises of "science," an objective value standard, one that "should" be imposed on all persons. Strictly interpreted, of course, almost any activity each of us undertakes is, in some way or another, a possible risk to our health. Once this is recognized, the question is one of drawing lines, and there is no well-defined set of activities that fall into one category or the other.

Toward a Sumptuary Constitution

We have been caught up in a wave of politicization for several decades. As a result, the set of activities that have been subjected to governmental-bureaucratic prohibition, regulation, and control has been expanded dramatically. Once politics was discovered as the apparent low-cost means of imposing preferences on behavior, a Pandora's box was opened that shows no signs of closing itself.

In these as in other aspects of the relationship between the citizens and the government, the dangers of excessive politicization cannot be avoided merely by a change in the makeup of political parties or by a change of politicians. In democracy, politicians respond to the electorates, and electoral majorities may, in a piecemeal fashion, close off one liberty after another. Prediction of such a prospect suggests that genuine reform can come only by *constitutional* rules that will prevent ordinary democratic majorities, in the electorates or in legislative assemblies, from entering too readily into the sumptuary areas of activities. Until and unless we recognize that politics, too, must operate within constitutional limits, each of our liberties, whether valued highly or slightly, is up for grabs.

Notes

1. See R.H. Coase, "The Problem of Social Cost," *Journal of Law and Economics* (October 1960): 1–44. Briefly stated, the Coase theorem is that efficient outcomes of interactions will emerge so long as persons are free to enter into voluntary contractual agreement. In the example, if I place a higher negative value on silence than the positive value that my neighbor places on rock music, I can successfully bribe him or her. Perhaps only to economists would this explicit approach to mutual

adjustment seem plausibly meaningful. Indirect exchanges, in behavioral rather than monetary dimensions, would be normal in many settings.

2. See Amartya K. Sen, "The Impossibility of a Paretian Liberal," *Journal of Political Economy* 78 (1970): 152–157.
3. For a more technical discussion of the potential externalities of political process, see my paper, "Politics, Policy, and the Pigovian Margins," *Economica* 29 (February 1962): 17–28.

13

Concluding Remarks

Robert D. Tollison

THE purpose of this book has been twofold. First, the Appendix, which follows this conclusion, surveys the existing evidence on the potential health impacts of ETS. As any fair-minded reader will conclude from reading the Appendix, a health case against ETS does not exist; this conclusion is based on a mountain of research. Second, the chapters in this book provide a cross-section of commentary on ETS, ranging from the viewpoints of scholars to those of prominent individuals who play active, day-to-day roles in the economy and society. Their views are different from what the lay reader normally hears about ETS. These writers stress common sense, accommodation, patience, scientific research, good manners, corporate and union responsibility, and various other similar metaphors for dealing with the social issues posed by ETS. Such viewpoints deserve to be heard in the debate about ETS, and if they are heard, this book will have fulfilled its purpose.

Moreover, if the advocacy of the hotheads on both sides of the issue can be avoided, rational discourse may follow. It is my view that most smoking behavior can be approached as a matter of private choice, not public choice. Entities in the private and public sectors can adopt the smoking policies that their workers and patrons prefer. This does not mean laissez-faire for smokers. Some entities will ban smoking, and some will not. Some entities will install better ventilation systems. Still others will segregate smokers and nonsmokers. And so on. The pattern that emerges across entities, however, will be the pattern preferred by smokers and nonsmokers alike. It will be the result of private alternatives and choices.

The public choice alternative involves having an administrative agency or a legislature choose a single rule for all to follow. The alternatives become simply smoking or no smoking rather than smoking or no smoking depending on the context. The private choice alternative is consistent with the practices of a free society. The public choice alternative is the route to a tyranny of personal preferences by one group over another. If the tyranny can be implemented over smokers, ask yourself who and what comes next.

Appendix
Research Evidence on Environmental Tobacco Smoke

The risk of reporting chronic cough, chronic wheeze, lower respiratory illness, bronchitis in the last year, chest illness in the last year and doctor diagnosed respiratory illness before age two was consistently higher among children living in homes positive in one or more of the moisture characteristics. These associations generally remained after adjustment for parental respiratory illness, current maternal smoking and parental education.

B. Brunekreef, D.W. Dockery, F.E. Speizer, J.H. Ware, J.D. Spengler, and B.G. Ferris, Jr. (Harvard School of Public Health and Harvard Medical School, Boston, and the Agricultural University of Wageningen, the Netherlands), "An Association between Moisture in the Home and Respiratory Symptoms in Primary School Children," *American Review of Respiratory Disease Supplement* 135 (4, pt. 2) (1987): A340 (abstract).

The criticisms of the Japanese (Hirayama) and Greek (Trichopoulos) studies were basically related to the general lack of reliability of questionnaires, inadequate histological diagnoses, uncertain accuracy of reported smoking habits, and failure to take into account general urban pollution and indoor pollutions emanating from cooking and heating practices.

W. Allan Crawford, (consultant in environmental and occupational health, Sydney, Australia), "Environmental Tobacco Smoke: The Use of Mathematical Models to Predict Health Effects," *Environment International* 13 (1987): 151–154.

Thus, the Repace and Lowrey tentative mathematical modelling with its preliminary results has as its basic input significantly flawed data with regard to environmental factors. It is based on epidemiologic data of a highly suspect nature which have been severely criticized by expert bodies. Their publications have engendered unwarranted anxiety in sections of the public who are concerned about the alleged health effects of environmental tobacco smoke.

Ibid.

The [National Academy of Sciences: *The Airliner Cabin Environment: Air Quality and Safety*] Committee's report provides no citation or data to support the statement that high transient concentrations of tobacco smoke components occur in the nonsmoking section of the cabin.

U.S. Department of Transportation and Related Agencies Appropriations Act, 1988, Speech of Hon. Tom DeLay of Texas, House of Representatives, July 13, 1987, in *Congressional Record,* July 15, 1987, pp. E2904–E2908.

It may be said that so far even toxicology has not been able to ascertain with any greater degree of probability than did epidemiology that there exists a link between damage to health and passive smoking.

G.J. Gostomzyk (director, Health Bureau, Augsburg, Germany), "Passive Smoking—Report on an International Symposium (23–25 October, 1986, Essen)," *Public Health* 49 (1987): 212–215.

The Surgeon General, like myself, is a politician. I have learned that when dealing with politicians it is better to watch what they do as opposed to listening to what they say. This is as true for the Surgeon General as it is for me. The simple fact of the matter is that the conclusions in the Surgeon General's report are not supported by the research in his own report.

Hon. Walter B. Jones of North Carolina, House of Representatives, February 18, 1987, "Inconclusive Evidence on the Harmful Effects of Smoking," *Congressional Record,* February 18, 1987, pp. E489–E490.

The present study supported the conclusions of Rogers et al. (1984) that exposure to cigarette smoke in the home was not a risk factor for middle ear problems in children. Rogers et al. (1984) reanalyzed previous data from Kraemer et al. (1983) by controlling for the influence of nasal congestion. They suggested that it was the nasal congestion that influenced Kraemer's significant results, not exposure to household cigarette smoking.

The relative elimination of air pollution in the present investigation suggested that tobacco smoke exposure alone might not be as great a risk factor for middle ear problems in children as previously assumed.

> Ken J. Kallail (University of Kansas School of Medicine, Wichita), Harry R. Rainbolt (Kansas State University), and Melvin D. Bruntzel (Kansas Department of Education), "Passive Smoking and Middle Ear Problems in Kansas Public School Children," *Journal of Communication Disorder* 10 (3) (1987): 187–196.

A total of 14 studies on lung cancer risk for the non-smoker in relation to spousal smoking survive critical assessment despite their technical flaws. Of the 6 studies based on American subjects, none shows a statistically significant increase in risk for the nonsmoker with a spouse who smokes.

Based on the evidence to date, the concern about the risk of lung cancer for nonsmoking Americans appears to be overstated and unsupported.

> Alan W. Katzenstein (Katzenstein Associates, Larchmont, New York), "Environmental Tobacco Smoke ETS and Risk of Lung Cancer—How Convincing Is the Evidence," Washington, D.C., Tobacco Institute, March 1987.

In summary, when a spouse's smoking status is used to estimate a nonsmoker's ETS exposure, "a considerable amount of misclassification" may result. Since selection bias and confounding must also be considered, extreme caution is required in the interpretation of these studies. This is especially so when the literature as a whole contains several studies reporting no significant association between ETS exposure and lung cancer as well as various inconsistencies, both among and within studies.

> S.J. Kilpatrick, Jr. (Medical College of Virginia, Virginia Commonwealth University, Richmond, Virginia), "Misclassification of Environmental Tobacco Smoke Exposure: Its Potential Influence on Studies of Environmental Tobacco Smoke and Lung Cancer," *Toxicology Letters* 35 (1987): 163–168.

Lifetime exposures to environmental tobacco smoke from the home or workplace for 88 "never-smoked" female lung cancer patients and 137 "never-smoked" district controls were estimated in Hong Kong to assess the possible causal relationship of passive smoking to lung cancer risk. Relative risks based on the husband's smoking habits, or lifetime estimates of total years, total hours, mean hours/day, or total cigarettes/day smoked by each household smoker did not show dose-response results. Similarly, when such categories as mean hours/day, or earlier age of initial exposure, were combined with years of exposure, there were no apparent increases in relative risk.

> Linda C. Koo, John H-C. Ho, Daisy Saw, and Ching-yee Ho (Department of Community Medicine, University of Hong Kong), "Measurements of Passive Smoking and Estimates of Lung Cancer Risk among Non-Smoking Chinese Females," *International Journal of Cancer* 39 (1987): 162–169.

Bias caused by misclassification of smoking habits coupled with between-spouse smoking habits concordance can completely explain reported apparent excesses in lung cancer risk in non-smokers married to smokers.

> Peter N. Lee, "Lung Cancer and Passive Smoking: Association an Artefact Due to Misclassification of Smoking Habits," *Toxicology Letters* 35 (1987): 157–162.

In the event, whether the true relative risk is 1.05 or 1.14, it is unlikely that any epidemiological study has been, or can be, conducted which could permit establishing that the risk of lung cancer has been raised by passive smoking. Whether or not the risk is raised remains to be taken as a matter of faith according to one's choice.

> Nathan Mantel (professor of mathematics, statistics, and computer science, American University, Bethesda, Maryland), "Lung Cancer and Passive Smoking," *British Medical Journal* 294 (1987): 440.

Other possible more serious biases in the studies conducted were not considered. (These include publishing bias: if an investigator got a weakly or insignificantly negative result for the role of passive smoking in lung cancer would he bother submitting it for publication? And if he did, would it be accepted? There seems to be a tendency toward accepting uncritically or less critically manuscripts which are on the right side of the fence on the issue of passive smoking.)

> Ibid.

Children living in damp houses, especially where fungal mold was present, had higher rates of respiratory symptoms, which were unrelated to smoking in the household, and higher rates of symptoms of infection and stress. Housing should remain an important public health issue, and the effects of damp warrant further investigation.

Claudia J. Martin, Stephen D. Platt, and Sonja M. Hunt (Research Unit in Health and Behavioural Change, University of Edinburgh), "Housing Conditions and Ill Health," *British Medical Journal* 294 (1987): 1125–1127.

A nonsmoking employee in a typical office would have to work more than six and a half 40-hour weeks in a row to be exposed to nicotine "equivalent" of one cigarette, according to a new scientific study released today.

The study also shows that it would take a marathon eating session lasting 398 continuous hours for a diner in a restaurant to be exposed to the nicotine "equivalent" of a single cigarette.

The study findings by an independent scientific team, cover actual scientific measurement of nicotine—an indicator of cigarette smoke in the air—in 38 offices and 36 restaurants randomly selected in Dallas.

"The levels of nicotine found in the study indicate that smoking regulations are unnecessary in order to assure adequate indoor air quality," said David Weeks, M.D., an expert on substances in indoor air.

Tobacco Institute, news release, *New Study of Actual Air Quality in Restaurants, Offices Shows Tobacco Smoke an Insignificant Factor* (June 30, 1987).

Research efforts on indoor air quality should consider fungi as an integral and important parameter to study.

Health and Welfare Canada Working Group on Fungi and Indoor Air, *Significance of Fungi in Indoor Air: Report of a Working Group*. Distributed at Indoor Air '87, the Fourth International Conference on Indoor Air Quality and Climate, West Berlin, August 1987.

Prohibition of smoking has not been shown to have any measurable effect on either indoor air quality or associated health and comfort symptoms of sick building syndrome. Ventilation required to remove indoor contaminants produced by the occupants themselves, specifically CO_2

and body odor, will also remove the constituents of ETS. On the other hand, if adequate ventilation rates are not provided, then indoor-generated substances and dusts and chemicals infiltrating the building envelope from outdoors increase in concentration to unacceptable levels, even if ETS should be entirely absent.

. . . In modern office buildings under normal use and occupancy, . . . ETS does not appear to contribute significantly to a build-up of contaminants in offices.

Theodore D. Sterling, Chris W. Collett, and Elia M. Sterling, "Environmental Tobacco Smoke and Indoor Air Quality in Modern Office Work Environments," *Journal of Occupational Medicine* 29 (1)(1987): 57–62.

None of the six case control studies yielding a positive relationship was conducted according to the state of art of epidemiological research, giving reasonable and sound evidence which cannot be explained by chance, bias, confounding or misclassification.

The available studies, while showing some evidence of association, do not exclude chance, bias or confounding. They provide, however, a serious hypothesis. Further studies are needed, if one wants to come to an adequate and scientifically sound conclusion concerning the question as to whether passive smoking causes lung cancer in man.

Of the three prospective studies, only one shows a moderate risk increase of 1.74.

Of the 12 case-control studies, two contribute nothing to the evidence, six show a moderate risk increase, but do not sufficiently exclude chance, bias and confounding, four studies show a moderate risk decrease or no risk change.

All studies with positive associations can just as well be explained by change, bias, confounding or misclassification. Such poorly conducted and inconclusive studies cannot be added or pooled to get convincing evidence, as has been attempted in serious efforts to evaluate the situation. Science should disregard poor studies. False plus false does not equal true.

K. Uberla, "Lung Cancer from Passive Smoking: Hypothesis or Convincing Evidence," *International Archives of Occupational and Environmental Health* 59 (1987): 421–437.

The FAA/PHS 1970–71 study of tobacco smoke contamination in aircraft has been labeled by some as "inadequate." However, the limited data that have been collected in subsequent less-structured testing, prior to the NAS study, have done little to refute the conclusion reached in the early study, i.e., that, based on environmental levels and expected dose-response relationships of contaminants, tobacco combustion products do not represent a hazard to the nonsmoking passengers.

> U.S. Department of Transportation, *Report to Congress Airline Cabin Air Quality,* Report of the Department of Transportation to the United States Congress (Washington, D.C.: Government Printing Office, February 1987).

The NAS Committee recommends a ban on smoking on all domestic commercial flights, for four major reasons: to lessen irritation and discomfort to passengers and crew, to reduce potential health hazards to cabin crew associated with ETS (environmental tobacco smoke), to eliminate the possibility of fires caused by cigarettes, and to bring the cabin air quality into line with established standards for other closed environments.

The Committee stated that "Empirical evidence is lacking in quality and quantity for a scientific evaluation of the quality of airliner cabin air or of the probable health effects of short or long exposure to it."

We agree that exposure to ETS could be viewed as a problem by some crew and passengers. However, we believe that further study is needed before the Department can propose a definitive response to this recommendation.

> Ibid.

None of the four epidemiological studies which specifically examine the effect of exposure of women to ETS at work finds a statistically significant increase in risk.

> Anthony Arundel, Ted Irwin, and Theodor Sterling (Faculty of Applied Sciences, School of Computing Science, Simon Fraser University, Burnaby, British Columbia, Canada), "Nonsmoker Lung Cancer Risks from Tobacco Smoke Exposure: An Evaluation of Repace and Lowrey's Phenomenological Model," *Journal of Environmental Science and Health* C4 (1) (1986): 93–118.

There are three problems with Repace and Lowrey's phenomenological model. (1) The lung cancer risk estimates derived from the SDA study are based on very few observed deaths and are unstable. (2) The apparent differences in lung cancer mortality between SDA and nonSDA never smokers may be due to a variety of factors other than ETS exposure. (3) A number of assumptions and calculations are made that are clearly incorrect.

For several age groups, the phenomenological model predicts more lung cancer deaths among never smokers as a result of exposure to ETS alone than a reasonable estimate of the total number of lung cancer deaths from all causes.

Repace and Lowrey's phenomenological estimate of the lung cancer risk for nonsmokers from ETS exposure is unstable and inaccurate. The evidence presented by Repace and Lowrey to support the plausibility of their estimate is based on several errors and unrealistic assumptions. Without corroborating data from independent sources, little confidence can be placed in their results.

Ibid.

Of the four studies that have considered ETS exposure outside the home, three have attempted to define some composite measure of exposure from several sources. None of these analyses yielded a significant relationship between composite ETS exposure and lung cancer. The case-control study of Garfinkel, et al., which represents the most rigorous effort to date to document total ETS exposure, found no dose-response relationship between the hours/day of ETS exposure from all sources and lung cancer risk.

The epidemiological data concerning any cause and effect relationship between ETS and lung cancer is, at best, equivocal. The most persistent weakness is the absence of adequate exposure information. This precludes the evaluation of dose-response relationships—a critical element of the test for causality. Moreover, problems of sampling bias and misclassification complicate interpretation of studies done to date.

Nancy J. Balter (Department of Biology, Georgetown University, Washington, D.C.), S. James Kilpatrick (Department of Biostatistics, Medical College of Virginia, Richmond, Virginia), Philip Witorsch (Department of Medicine, George Washington University, Washington, D.C.), and Sorell L. Schwartz (Department of Pharmacology, Georgetown University, Washington, D.C.), "Causal Relationship between Environmental Tobacco Smoke and Lung Cancer in Non-smokers: A Critical Review of the Literature," Proceedings of the Air Pollution Control Association, April 11, 1986.

Risk extrapolation based on the presence of carcinogens, in the absence of appropriate epidemiological data, does not *establish* causation. It *presumes* causation.

Ibid.

The conclusion that exposure to ambient tobacco smoke produces ". . . about 5,000 lung cancer deaths per year in U.S. nonsmokers aged 35 years . . ." belongs more to speculation than reality. When we take into account the outcome of the randomized trials together with the paradoxical implications of the findings of Sandler et al. (1985), the best estimate of LCDs per year is approximately zero.

P.R.J. Burch (Department of Medical Physics, University of Leeds, General Infirmary, Leeds, United Kingdom), "Health Risks of Passive Smoking: Problems of Interpretation," *Environment International* 12 (1986): 23–28.

A total of 16 scientific contributions and working papers were presented at the Vienna meeting that closed with a round-table discussion during which important aspects of the symposium were gone into once more, in greater depth. With the exception of Dr. Hirayama, all the participants agreed that so far there was no definite proof of a causal relationship between passive smoking and the risk of lung cancer. None of the epidemiological studies considered satisfied the essential criteria of scientific methodology.

Joseph Handler (former director of the World Health Organization), Preface to *Passiverauchen aus medizinischer Sicht [Medical perspectives on passive Smoking]* (Geneva: IRL Imprimeries Reunies Lausanne, 1986), p. A.

The question as to whether the conceivable theoretical possibility of risk calls for official preventive measures is not a medical but a politico-social one.

Ibid., p. B.

After a detailed scientific evaluation of all the major buildings we have studied, we have determined that high levels of environmental tobacco smoke were the medical cause of indoor pollution problems in only 4 percent of those buildings.

This result has also been corroborated by the work of NIOSH. In a similar study of 203 buildings over a 5-year period, they found that environmental tobacco smoke was the cause of problems in only 2 percent of their buildings.

It is also significant that in every single building we found environmental tobacco smoke had accumulated we also found that high levels of fungal spores were also present in the air inside those buildings.

The reason environmental tobacco smoke often takes the blame for these symptoms is obvious: it is the only visible indoor pollutant. However, we have determined that the presence of high concentrations of tobacco smoke indicate the much more serious problem of poorly designed and an improperly maintained ventilation system.

> U.S. Congress, House of Representatives, Hearings before the Subcommittee on Health and the Environment of the Committee on Energy and Commerce, Gray Robertson, statement, *Designation of Smoking Areas in Federal Buildings,* 99th Cong., 2d sess., June 12, 27, 1986.

The lung function of non-smokers who lived in homes with (a) no smokers, (b) a smoker who smoked one pack per day, and (c) a smoker who smoked two or more packs per day, were compared. In this comparison all members of the household were included and this total group was separated into those who lived in homes with electric stoves and those who lived in homes with gas stoves. In general, the lung function of non-smokers who lived in homes with electric stoves was not influenced by passive smoke.

The lung function of non-smokers who lived in homes with gas stoves did not appear to be affected by passive smoke, but the numbers in each smoking category was small.

> H. Roland Hosein and Paul Corey (Occupational and Environmental Health Unit, University of Toronto), "Domestic Air Pollution and Respiratory Function in a Group of Housewives," *Canadian Journal of Public Health* 77 (1986): 44–50.

Our examination has shown that the figures derived by the authors [Repace and Lowrey] are based upon incorrect theoretical assumptions and inflated empirical estimates. The other calculations, contained in their paper, suggest to us that the tendency to choose inflated estimates with regard to exposure in the home was consistently followed. As a result, one must conclude that the estimate of 5,000 lung cancer deaths per year in the United States, due to exposure to ambient tobacco smoke, does not represent an accurate assessment of the problem.

Clark Johnson (Heinz Letzel, Munich, Germany), "Letter to the Editors," *Environment International* 12 (1986): 21–22.

In the examination of the relationship between levels of air pollutants and respiratory health, it is very important that any confounding effect of covariables be distinguished from the effect of air pollution itself. We have shown that the industrial area, which has the highest level of [total suspended particulates], has also the highest prevalence of domestic smoking, parental respiratory symptoms, and gas cooking.

Anthony T. Kerigan, Charles H. Goldsmith, and L. David Pengelly, "A Three-Year Cohort Study of the Role of Environmental Factors in the Respiratory Health of Children in Hamilton, Ontario. Epidemiologic Survey Design, Methods, and Description of Cohort," *American Review of Respiratory Disease* 133 (1986): 987–993.

The null hypothesis that passive smoking and lung cancer mortality are causally unrelated still stands. Until it is rejected, I consider it irresponsible to apply risk management techniques, even if they had been applied correctly.

S. James Kilpatrick (Medical College of Virginia, Richmond, Virginia) "Letters to the Editors," *Environment International* 12 (1986): 29–31.

The potential association of lung cancer with passive smoking has been studied epidemiologically, clinically, and with mathematical models. There have been both positive and negative studies, such that the association is still considered a potential rather than probable one.

Michael D. Lebowitz (professor, Division of Respiratory Sciences, University of Arizona College of Medicine, Tucson, Arizona), "The Potential Association of Lung Cancer with Passive Smoking," *Environment International* 12 (1986): 3–9.

Only mathematical models, such as that of Repace and Lowrey (1985), appear to estimate dosage to humans from environmental tobacco smoke that may be carcinogenic. These models are based on mathematical assumptions and calculations that appear to include overestimations. The dose so derived does not reflect estimates based on actual controlled laboratory studies. Actual estimates imply a dosage to the passive smoker of less than 2 cigarettes/day, except in unusual circumstances; these estimates are ten- to one-hundred-fold less than that in the Repace and Lowrey model.

The model and methods used by Repace and Lowry (1985) and others do not provide estimates of precision, confidence intervals, or consistency estimates, the very statistical bases of estimation procedures.

Ibid.

If misclassification of smokers and non-smokers and other possible biases are not taken into account the result is likely to be an estimate of lung cancer deaths attributed to passive smoking that is incompatible with the amounts of smoke to which non-smokers are exposed.

Peter N. Lee, "Misclassification as a Factor in Passive Smoking Risk," *Lancet* 2(8511) (1986): 867.

Amongst lifelong non-smokers, passive smoking was not associated with any significant increase in risk of lung cancer, chronic bronchitis, ischaemic heart disease or stroke in any analysis. From all the available evidence, it appears that any effect of passive smoke or risk of any of the major diseases that have been associated with active smoking is at most small, and may not exist at all.

P.N. Lee, J. Chamberlain, and M.R. Alderson (Institute of Cancer Research, United Kingdom), "Relationship of Passive Smoking to Risk of Lung Cancer and Other Smoking-Associated Diseases," *British Journal of Cancer* 54 (1986): 97–105.

Our analyses showed no significant effect of passive smoking on lifelong non-smokers as regards risk of chronic bronchitis, ischaemic heart disease or stroke. In all the analyses relating the various indices of passive smoke exposure to these diseases, no significant differences were seen and slight decreases in risk were as common as slight increases.

Ibid.

Overall the results showed no evidence of an effect of passive smoking on lung cancer incidence among lifelong non-smokers. In male patients, relative risks were increased from some of the indices but numbers of cases were small and none of the differences approached statistical significance. In females, where numbers of cases were larger, such trends as existed tended to be negative and indeed were marginally significantly negative for passive smoking during travel and during leisure. For the combined sexes no differences or trends were statistically significant at the 95% confidence level; such trends as existed tending to be slightly negative.

Ibid.

It is not known if any tobacco incineration products per se are immunogenic in man.

To date, the role of hypersensitivity is as yet undefined. In spite of the considerable number of previous studies and their supporting evidence of an allergic response to tobacco smoke, the obvious pitfalls of these studies (undefined study populations, inappropriate antigens, and inadequate documentation of smoke sensitivity) undermine any significant conclusions.

Samuel B. Lehrer, Richard P. Stankus, and John E. Salvaggio (Clinical Immunology Section, Department of Medicine, Tulane Medical Center, New Orleans), "Tobacco Smoke Sensitivity: A Result of Allergy?" *Annals of Allergy* 56 (1986): 1–10.

With social class not allowed for in the analysis [regarding parent's smoking], any social class effect can readily be misinterpreted as a passive smoking effect, inasmuch as the social classes differ so much in their smoking rates.

N. Mantel, "Letter to the Editor: Does Passive Smoking Stunt the Growth of Children," *International Journal of Epidemiology* 15(3) (1986): 427–428.

Three studies have shown a small reduction in pulmonary function in normal adults exposed to ETS. Interpretation of these findings is difficult because pulmonary effects in normal adults are likely to reflect the cumulative burden of many environmental and occupational exposures and other insults to the lung. Thus, the effects of ETS on the lungs of adults are likely to be confounded by many other factors, making it difficult to attribute any portion of the effect solely to ETS.

National Academy of Sciences, *Environmental Tobacco Smoke: Measuring Exposures and Assessing Health Effects* (Washington, D.C.: National Academy Press, 1986), p. 10.

Reports have noted an excess risk of cardiovascular disease in ETS-exposed nonsmokers; however, methodological problems in the designs and analyses of these studies preclude any firm conclusions about the results. Studies reporting that ETS can precipitate the onset of angina pectoris among people who already have this condition are subject to the same precautionary note. Exposure to ETS produced no statistically significant effects on heart rate or blood pressure in school age children or healthy adult subjects, either during exercise or at rest. Data are not available as to possible adverse cardiovascular effects in susceptible populations, such as infants, elderly, or diseased individuals.

Ibid.

A study released today shows that a passenger seated in a no-smoking section of a U.S. commercial airliner would have to complete eight continuous New York-to-Tokyo round trips to be exposed to the nicotine equivalent of one cigarette.

The amount of cigarette smoke in the no-smoking sections of commercial airline cabins is so small that it would take the equivalent of 224 hours, or more than nine days, of non-stop flying to reach the exposure level, according to Guy Oldaker, senior research chemist of the R.J. Reynolds Tobacco Company.

All four systematic in-air studies—including a 1971 report from the Department of Health, Education and Welfare and the Federal Aviation Administration, a 1983 study by a team from San Francisco General Hospital Medical Center and a 1984 study by Japanese researchers—reached the conclusion that the amount of cigarette smoke present in actual commercial flights is extremely small, and that the levels do not indicate a demonstrated risk to passengers or flight personnel.

Tobacco Institute, news release, "Study Shows Additional Airline Smoking Curb Proposed by National Academy of Sciences Is Not Justified" (August 12, 1986).

A 1985 poll by Tarrance and Associates of a representative sample of 1,000 frequent flyers found that 83 percent—including 79 percent of ex-smokers and 81 percent of those who have never smoked—support the current smoking rules on commercial airlines. Department of Transportation (DOT) complaint files indicate only 2–3 percent of complaints from airline passengers have to do with smoking.

Ibid.

In this population, highly significant associations emerged between damp and mold in the house and respiratory morbidity in children, at least as reported by their parents.

Parental smoking emerged as a less significant factor than might have been supposed, but the analysis excluded the possibility that either smoking or gas fumes could account for the observed association between damp, moldy housing and lower respiratory morbidity in this sample of primary schoolchildren.

> David P. Strachan and Robert A. Elton, "Relationship between Respiratory Morbidity in Children and the Home Environment," *Family Practice* 3(3) (1986): 137–142.

Further studies on the relationship between involuntary smoking and cardiovascular disease are needed in order to determine whether involuntary smoking increases the risk of cardiovascular disease.

> *Surgeon General's Report* (Washington, D.C.: Government Printing Office, 1986), p. 14.

The physiologic and clinical significance of the small changes in pulmonary function found in some studies of adults remains to be determined. The small magnitude of effect implies that a previously healthy individual would not develop chronic lung disease solely on the basis of involuntary tobacco smoke exposure in adult life.

> Ibid., p. 62.

There are no studies of acute respiratory illness experience in adults exposed to environmental cigarette smoke.

> Ibid.

Several conclusions arise from this study. Episodes of lower respiratory illness, defined as those in which there were one or more consultations at which adventitious lung sounds were recorded, are particularly frequent in the children of manual workers. This cannot be explained by the many social and family variables examined in this study such as overcrowding, smoking habits, parents' respiratory symptoms, and breast feeding.

> C.J. Watkins, Y. Sittampalam, D.C. Morrell, S.R. Leeder, and E. Tritton (Department of General Practice, United Medical and Dental Schools of

Guy's and St. Thomas's Hospitals, London) "Patterns of Respiratory Illness in the First Year of Life," *British Medical Journal* 293 (1986): 794–796.

We studied the acute effects of one hour of passive cigarette smoking on the lung function and airway reactivity of nine young adult asthmatic volunteers. Passive smoking produced no change in expiratory flow rates. We conclude that passive smoking presents no acute respiratory risk to young asymptomatic asthmatic patients.

Herbert P. Wiedemann, Donald A. Mahler, Jacob Loke, James A. Virgulto, Peter Snyder, and Richard A. Matthay (Department of Medicine, Yale University School of Medicine, New Haven), "Acute Effects of Passive Smoking on Lung Function and Airway Reactivity in Asthmatic Subjects," *Chest* 89(2) (1986): 180–185.

A thorough and critical examination of the relevant literature fails to provide compelling evidence that exposure to ambient tobacco smoke produces adverse chronic health effects.

P. Witorsch (physician and professor of medicine, George Washington University, Washington, D.C.), "Passive Smoking," *New Zealand Medical Journal* November 12, 1986, p. 865.

A review of the literature indicates that there is no substantial evidence to support the view that exposure to environmental tobacco smoke presents a significant health hazard to the nonsmoker.

D. Aviado (physician and former professor of medicine) "Health Issues Relating to 'Passive Smoking,' " in R.D. Tollison, ed., *Smoking and Society* (Lexington, Mass.: Lexington Books, 1985), p. 158.

The results suggest that many people, both smokers and nonsmokers, may be at risk from CO generated by certain domestic heating systems and that nonsmokers are far more likely to be exposed to high levels of CO from these sources than from being in a room with a heavy smoker. Poor ventilation associated with the current trend towards excluding all draughts is likely to exacerbate the situation for both smokers and nonsmokers.

B.D. Cox and Margaret J. Whichelow (Office of the Regius Professor of Physic, Cambridge University School of Clinical Medicine, Cambridge), "Carbon Monoxide Levels in the Breath of Smokers and Nonsmokers: Effect of Domestic Heating Systems," *Journal of Epidemiology and Community Health* 39 (1985): 75–78.

Other covariates had virtually no explanatory power, and were rather unstable across subpopulations: educational levels of head of household, sex, and mother's smoking status. In particular, we found that a mother's smoking in the home was unrelated to acute respiratory disease incidence of her children. However, this should not be too surprising in view of the contradictory findings on the health effects of passive smoking.

Winston Harrington and Alan J. Krupnick, "Short-Term Nitrogen Dioxide Exposure and Acute Respiratory Disease in Children," *Journal of the Air Pollution Control Association* 35 (1985): 1061–1067.

There were no significantly increased risks for having a mother, a father, or spouse(s) who smoked or for being exposed at work.

Our data are not consistent with the findings with regard to nonsmokers obtained by Hirayama and Trichopoulos et al.

The association of lung cancer risk with exposure to coal heating or cooking warrants further investigation.

Anna H. Wu, Brian E. Henderson, Malcolm C. Pike, and Mimi C. Yu (Department of Family and Preventive Medicine, University of Southern California School of Medicine), "Smoking and Other Risk Factors for Lung Cancer in Women," *Journal of the National Cancer Institute* 74(4) (1985): 747–751.

Children exposed to air pollution in the home from a cigarette-smoking parent show no adverse lung performance, a University of Toronto specialist said Tuesday.

Roland Hosein told an environmental concerns conference that while he cannot fully explain it, tests in children exposed to cigarette smoke alone showed no deterioration in their lung function as measured by their ability and rate of exhaling.

"Conference Told Smoke Didn't Hurt Children's Lungs," *Regina Leader-Post,* September 27, 1984, B10.

Because of concern among flight attendants about passive exposure to cigarette smoke during work on commercial aircraft, a preliminary investigation was conducted to search for an increase in expired air (end tidal) carbon monoxide in flight attendants after work.

All flights were "turnaround" flights from Los Angeles to Honolulu and back. These flights were of about five hours' duration in each direction.

There was no increase in the concentration of carbon monoxide in the expired air (end tidal) of these flight attendants during the flights in this study. In fact, their exhaled air carbon monoxide levels decreased by an insignificant amount.

These results indicate that the concentration of smoke to which flight attendants are passively exposed is too low to alter significantly their expired air carbon monoxide levels.

> Douglas B. Duncan and Peter P. Greaney (University of California, Irvine, Southern Occupational Health Center), "Passive Smoking and Uptake of Carbon Monoxide in Flight Attendants," *Journal of the American Medical Association* 251(20) (1984).

No convincing differences for viral infection or respiratory illness were seen with parental smoking as an isolated factor.

> Gregory Gardner, Arthur L. Frank, and Larry H. Taber (Influenza Research Center, Department of Microbiology and Immunology, and Baylor College of Medicine, Houston, Texas), "Effects of Social and Family Factors on Viral Respiratory Infection and Illness in the First Year of Life," *Journal of Epidemiology and Community Health* 38 (1984): 42–48.

Taking our own results into consideration, it seems that the passive inhalation of tobacco smoke at home or in the workplace by healthy individuals probably does not lead to any essential impairment of pulmonary function.

> Michael Kentner, Gerhard Triebig, and Dieter Weltle (Institute of Occupational and Social Medicine and Polyclinic of Occupational Diseases of the University of Erlangen-Nuremberg), "The Influence of Passive Smoking on Pulmonary Function—A Study of 1,351 Office Workers," *Preventive Medicine* 13 (1984): 656–669.

200 female lung cancer patients and 200 healthy district controls were interviewed to identify and quantify the various sources of passive smoking among Chinese females in Hong Kong. For the ever-smokers, passive exposure from external sources do not appear to add to their risk. For the never-smokers, qualitative assessments (smoke exposure categories, age

when passive exposure started), and quantitative assessments (hours, years, intensity) showed no significant differences between the data for patients and controls. Moreover, higher relative risks were not associated with higher levels of passive smoking for the ever- or never-smokers. Thus, our findings would seem to indicate that passive smoking, as an isolated factor, did not have an influence on female lung cancer incidence in Hong Kong.

> L.C. Koo, J.H-C. Ho, and D. Saw (Department of Community Medicine, University of Hong Kong, Hong Kong), "Is Passive Smoking an Added Risk Factor for Lung Cancer in Chinese Women?" *Journal of Experimental and Clinical Cancer Research* 3(3) (1984): 277–283.

We did not find any significant interaction between the smoking habits of either parent smoking and their spouses' lung function, similar to Comstock and coworkers and Schilling and Associates, but different from Kauffmann and coworkers.

> Michael D. Lebowitz, R.J. Knudson, and B. Burrow (University of Arizona College of Medicine, Tucson, Arizona), "Family Aggregation of Pulmonary Function Measurements," *American Review of Respiratory Disease,* 129 (1984): 8–11.

White and Froeb studied self-selected volunteers with regard to workplace exposure and reported some effects of passive smoking in a subset of their volunteers. . . . Even with a biased population, poor study design, and incorrect statistical evaluation, there were no clearcut, consistent, medically meaningful differences between passive smokers and groups of nonsmokers; a corrected statistical analysis strengthened this conclusion.

> Michael D. Lebowitz, "Influence of Passive Smoking on Pulmonary Function: A Survey," *Preventive Medicine* 13 (1984): 645–655.

From a few studies on occupational groups exposed to carbon monoxide (CO) and from experiments with animals chronically treated with CO or nicotine, the conclusion can be drawn that neither CO nor nicotine is likely to play a role in the development and progression of coronary heart disease in those concentrations normally found in passive smokers.

> H. Schievelbein and F. Richter (Institute of Clinical Chemistry, German Heart Center, West Germany), "The Influence of Passive Smoking on the Cardiovascular System," *Preventive Medicine* 13 (1984): 626–644.

Should lawmakers wish to take legislative measures with regard to environmental tobacco smoke, they will, for the present, not be able to base their efforts on a demonstrated health hazard from environmental tobacco smoke.

> H. Valentin (Executive Committee, German Society of Occupational Medicine) and E. Wynder (director, American Health Foundation), "Health Danger through Passive Smoking Not Proven: Physicians' View on Passive Smoking," press release from International Symposium, "Passive Smoking from a Medical Point of View," April 9–12, 1984, Vienna, Austria.

Passive smoking was not associated with an increase in total mortality.

> J.P. Vanderbroucke (epidemiologist, Department of Epidemiology, Erasmus University, the Netherlands), J.H.H. Verheesen, A. De Bruin, B.J. Mauritz (students, Department of Environmental and Tropical Health, Agricultural University, Wageningen), C. Van Der Heide-Wessel (retired community physician, Department of Health, City of Amsterdam), and R.M. van Der Heide (infectious disease physician, Academic Hospital, University of Amsterdam), "Active and Passive Smoking in Married Couples: Results of 25 Year Follow Up," *British Medical Journal* 288 (1984): 1801–1802.

We conclude that there is passive absorption of nicotine from tobacco smoke by flight attendants during a transoceanic flight but that the quantity consumed (equivalent to one cigarette) is relatively small compared with that consumed by cigarette smokers, and the concentrations achieved are unlikely to have physiologic effects.

> Donna Foliart, Neal L. Benowitz, and Charles E. Becker (San Francisco General Hospital Medical Center), "Passive Absorption of Nicotine in Airline Flight Attendants," *New England Journal of Medicine* 308(18) (1983): 1105.

Based on theoretical and empirical results, CO sidestream emissions from cigarettes have often been overemphasized.

> John R. Girman and Greg W. Traynor (staff scientists, Building Ventilation and Indoor Air Quality Program, Lawrence Berkeley Laboratory), "Indoor Concentrations," *Journal of the Air Pollution Control Association* 33(2) (1983): 90.

There were no significant changes in the pulmonary function parameters measured in any of the subjects when compared with baseline values.

Thus, passive exposure to cigarette smoke in these subjects produced marked symptoms described as usual asthma but not significant objective evidence of airways obstruction.

Alvin J. Ing and B.X. Breslin (Chest Unit, Concord Hospital, Sydney, Australia), "The Effect of Passive Cigarette Smoking on Asthmatic Patients." Paper presented at the Annual Scientific Meeting of the Thoracic Society of Australia, May 1983, pp. 541–543.

The present study showed no significant effect on $FEV_{1.0}$ from exposure to smokers in the home.

Jeffrey R. Jones, Ian T.T. Higgins, Millicent W. Higgins, and Jacob B. Keller (University of Michigan, School of Public Health, Department of Epidemiology, Ann Arbor), "Effects of Cooking Fuels on Lung Function in Nonsmoking Women," *Archives of Environmental Health* 38(4) (1983): 219–222.

There were no effects of ETS on [pulmonary function] or symptoms in children or adults, asthmatics or others.

Michael D. Lebowitz, "The Effects of Environmental Tobacco Smoke Exposure and Gas Stoves on Daily Peak Flow Rates in Asthmatic and Non-Asthmatic Families," in *ETS—Environmental Tobacco Smoke. Report from a Workshop on Effects and Exposure Levels,* ed. Ragnar Rylander et al. (Geneva: University of Geneva, March 15–17, 1983).

We measured oxygen uptake at the onset of pain in patients with stable angina pectoris, and found it to be reproducible.

There was no change in the mean oxygen uptake at the onset of angina with any intervention.

These results are in contrast with previous reports of the effects of smoking and carbon monoxide on exercise performance in angina pectoris.

M.W. McNicol and J.A. MCM Turner (Central Middlesex Hospital, London), "Oxygen Uptake at the Onset of Angina Pectoris: Effects of Nicotine and Carbon Monoxide," *Clinical Science* 65(3) (1983): 24.

A scientific study that the Environmental Protection Agency used to set a major air-quality standard was seriously flawed, and there is "considerable concern about the validity of the results reported."

At issue is the work of Dr. Wilbert S. Aronow, whose research into the pollutant's effect on heart patients was instrumental in setting the current standard. The EPA formed a team of government and outside scientists to look at Aronow's work last April, after the Food and Drug Administration found he had falsified data on an experimental heart drug.

In their report, released yesterday, the scientists said they "could not resolve the issue of possible falsification of data. However, we had considerable concern about the validity of the results reported. Raw data were lost or discarded, adequate records were not maintained, available data were of poor quality, quality control was nonexistent or inadequate. . . ."

"We conclude that EPA cannot rely on Dr. A's data due to the concerns we have noted," the report said.

> Cass Peterson, "EPA Probe Criticizes a Study Used in Air-Quality Standard," *Washington Post*, June 7, 1983, p. A15.

A review of the data from the studies which have been carried out or are in progress which address the effect of passive smoking on the respiratory system suggests that the effect varies from negligible to quite small. From this review, it was not possible to determine whether there is a specific group which is at increased risk or what the mechanism of the effect (if any) may be.

> U.S. Department of Health and Human Services, *Report of Workshop on Respiratory Effects of Involuntary Smoke Exposure: Epidemiologic Studies May 1–3 1983*, Public Health Service, National Institutes of Health (Washington, D.C.: Government Printing Office, December 1983).

I will agree that contributions to indoor levels of CO by smoking in "normally occupied residences" are probably quite small.

> John E. Yocom (TRC Environmental Consultants, Inc., East Hartford, Connecticut), "Indoor Concentrations. Author's Reply," *Journal of the Air Pollution Control Association* 33(2) (1983): 89.

There are less passive smokers among patients than the controls; and more non-smoking patients have non-smoking spouses. This finding is at variance with that of Dr. Hirayama's (1981).

> W.C. Chan and S.C. Fung, "Lung Cancer in Non-Smokers in Hong Kong," in *Cancer Epidemiology*, ed. E. Grundmann (New York: Gustav Fischer Verlag, 1982).

The pulmonary function testing showed that neither parental smoking nor gas home cooking fuel adversely affected lung function or yearly lung growth.

The results of the present study are good evidence that these factors do not affect the lung function of children living in the southwestern United States.

> Russell Dodge (assistant professor of medicine, University of Arizona, Tucson), "The Effects of Indoor Pollution on Arizona Children," *Archives of Environmental Health* 37(3) (1982): 151–155.

Given these and other criticisms of the White-Froeb study, it would appear that the New England Journal of Medicine has, perhaps unwittingly, performed a disservice to its readership. It is extremely unfortunate that a study so fraught with methodological problems, as indicated through numerous criticisms by scientists in the United States and elsewhere, should have been published in such a reputable journal of medicine. The White-Froeb study should, therefore, not be relied upon by the Congress, Federal agencies or other legislative or policymaking bodies when considering restrictions on smoking in public places.

> Hon. L.H. Fountain of North Carolina, House of Representatives, December 16, 1982, "White-Froeb Study Discredited by Scientists," *Congressional Record,* December 16, 1982, pp. E5252–E5254.

It must be concluded that passive smoking in the family, usually due to parental smoking habits, does not seriously affect permanent markers of respiratory disease such as pulmonary function.

> Michael D. Lebowitz, David B. Armet, and Ronald Knudson (Division of Respiratory Sciences, Westend Research Laboratories, Arizona Health Sciences Center, Tucson, Arizona), "The Effect of Passive Smoking on Pulmonary Function in Children," *Environment International* 8 (1982): 371–373.

The 1979 U.S. Surgeon-General's Report (US Public Health Service, 1979) devoted a chapter to the subject of allergy and tobacco smoke. It concluded that the existence of such an allergy was not clearly established but that those with a history of allergies to other substances, especially those with rhinitis or asthma, were more likely to report the irritating effects of tobacco smoke. Whether this was psychological, rather than a physiological, response is open to question.

> P.N. Lee (independent consultant in statistics, Surrey, England), "Passive Smoking," *Food and Chemical Toxicology* 20 (1982):223–229.

It is difficult to imagine that enclosure in a very smokey room did not have some emotional impact upon patients who were liable to angina, and the psychological disturbance may have done more to hasten the onset of symptoms than the increase of blood COHb.

R. Shephard, *The Risk of Passive Smoking* (London: Croom-Helm Ltd., Pub., 1982), p. 73.

The frequency of major respiratory symptoms among subjects showed little evidence of an association with the presence of some one else in the household who smoked cigarettes.

George W. Comstock, Mary B. Meyer, Knud J. Helsing, and Melvyn S. Tockman, "Respiratory Effects of Household Exposures to Tobacco Smoke and Gas Cooking," *American Review of Respiratory Disease* 124 (1981): 143–148.

The methods used by White and Froeb are open to criticism.

Franz Adlkofer, Gerhard Scherer, and H. Weimann (Forschungsgesellschaft Rauchen und Gesundheit mbH, Hamburg, Federal Republic of Germany), "Small-Airways Dysfunction in Passive Smokers," *New England Journal of Medicine* 303(7) (1980): 392–393.

Suspicions of such an allergy against tobacco smoke have existed for a long time, but have never been substantiated.

For present, the question as to whether allergy to cigarette smoke exists or not should be kept open.

Gunnar Bylin (Allergy Section, Department of Internal Medicine, Huddinge, Sweden), "Tobacco Allergy—Does It Exist?" *Lakartidningen* 77(16) (1980): 1530–1532 (translation).

White and Froeb do not have reliable estimates of the smoke exposure in the environment of their nonsmokers.

Gary L. Huber (Harvard Medical School, Boston), "Small-Airways Dysfunction in Passive Smokers," *New England Journal of Medicine* 303(7) (1980): 392–393.

Our data thus do not suggest that asthmatic subjects have an unusual sensitivity to cigarette smoke.

We would thus conclude that the specific sensitivity of asthmatic subjects is not a major consideration when determining air quality criteria for rooms contaminated by cigarette smoke.

Roy J. Shephard, R. Collins, and F. Silverman (Department of Preventive Medicine and Biostatistics, University of Toronto and the Gage Research Institute), " 'Passive' Exposure of Asthmatic Subjects to Cigarette Smoke," *Environmental Research* 20 (1979): 392–402.

The Aronow study must be evaluated in light of the fact that the endpoint of the study was highly subjective, that the stress factor was not controlled, and that a sham smoke or other environmental impingement was not used. In other words, not only was the sample small, but the scientific design was exceedingly poor.

E. Fisher, statement, U.S., Congress, House, Committee on Agriculture, Subcommittee on Tobacco, *Effect of Smoking on Nonsmokers,* Hearing, 95th Cong., 2d sess., September 7, 1978, pp. 35–40.

The influence of parents' tobacco smoking on the morbidity of the youngest children of 1–4 years of age, found by D.J. Hammer et al., could not be confirmed in our study.

Bo Holma, Ole-Winding (Institute of Hygiene, University of Copenhagen), "Housing, Hygiene, and Health. A Study in Old Residential Areas in Copenhagen," *Archives of Environmental Health* (1977): 86–93.

In accordance with the findings reported by Colley (14,15) and unlike those reported by a number of other investigators, there was no evidence that so-called passive smoking due to parental smoking had any appreciable effect on the appearance of respiratory symptoms in schoolchildren.

There is a definite relationship between respiratory symptoms in parents and those in children.

K.F. Kerrebijn, H.C.A. Hoogeveen-Schroot, and M.C. van der Wal, "Chronic Nonspecific Respiratory Disease in Children, A Five Year Follow-Up Study," *Acta Paediatrica Scandinavica* 261 (Suppl.) (1977): 7–72.

Respiratory symptoms, disease and lung function were studied in 376 families with 816 children who participated in a survey in three USA towns. Parental smoking had no effect on children's symptoms and lung function. Also, there was no evidence that passive smoking affected either lung function or symptoms of adults.

We conclude that exposure to low levels of smoke produced by cigarette smokers does not result in chronic respiratory symptoms or loss of lung function among children or among adults.

R.S.F. Schilling, A.D. Letai, S.L. Hui, G.J. Beck, J.B. Schoenberg, and A. Bouhuys, "Lung Function, Respiratory Disease, and Smoking in Families," *American Journal of Epidemiology* 106(4) (1977): 274–283.

In general, pollutant exposure was similar for control and disease subjects. The significant difference observed for particulate exposure was due to domestic smoking and was independent of the presence or absence of respiratory disease. We found no evidence to support a cause and effect relationship between pollutant exposure and respiratory disease.

Ralph E. Binder, Charles A. Mitchell, H. Roland Hosein, and Arend Bouhuys (Yale University Lung Research Center, New Haven), "Importance of the Indoor Environment in Air Pollution Exposure," *Archives of Environmental Health* 31(6) (1976): 277–279.

A study of the effects of family smoking habits on the symptoms of other family members has shown that symptoms of household members, especially children, are related to smoking habits within the households but are not significantly so when symptoms in adults are controlled.

Michael D. Lebowitz and Benjamin Burrows, "Respiratory Symptoms Related to Smoking Habits of Family Adults," *Chest* 69 (1976): 48–50.

A close association was found between parents' and children's respiratory symptoms that was independent of parents' smoking habits. There was no suggestion that exposure to the cigarette smoke generated when parents smoked had any more than a small effect upon the child's respiratory symptoms. While the sharing of genetic susceptibility between parents and children is a factor, therefore, cross infection, particularly in the families where parents smoke, is an important element in the association.

J.R.T. Colley, "Respiratory Symptoms in Children and Parental Smoking and Phlegm Production," *British Medical Journal* 2 (1974): 201–204.

About the Contributors

Peter L. Berger was born in Vienna, Austria. He attended Wagner College, where he received his B.A. in philosophy in 1949. He received his Ph.D. in sociology from New School for Social Research in 1954. Dr. Berger has taught at Evangelical Academy, Bad Boll, West Germany, University of North Carolina, Hartford Theological Seminary, New School for Social Research, Rutgers University, Boston College, Boston University, and is currently the director of the Institute for the Study of Economic Culture at Boston University. He has published ten major books and has been honored with a doctor of laws, Loyola University; doctor of humane letters, Wagner College; doctor of laws, University of Notre Dame; and as a U.S. representative, United Nations Working Group on the Right to Development.

Jock Bruce-Gardyne (Lord Bruce-Gardyne of Kirkden) was born in 1930 and educated at Winchester College and Magdalen College, Oxford. In his political career he has served as a member of Parliament (Conservative) for South Angulus (1964–74) and Knutsford (1979–83); parliamentary private secretary to the secretary of state, Scotland (1970–72); vice chairman, Conservative Parliamentary Committee (1972–74, 1980–81); minister of state treasurer, 1981; economic secretary to the treasurer, 1981–83, with responsibility for indirect taxation, including excise duties and monetary policy. He is the author of *Whatever Happened to the Quiet Revolution?: A Study of the Heath Government, 1970–1974* (1974), *The Power Game: A Study of Decision-making in British Government* with Nigel Lawson (1976) and *Ministers and Mandarins: A Study of the Relationship Between Politicians and Servants in Whitehall* (1986).

James M. Buchanan is University Distinguished Professor and general director of the Center for Study of Public Choice at George Mason University, Fairfax, Virginia. He is regarded as a founder of the public choice movement and has had a huge impact on the scope and direction of modern economics. Among his better-known works are ***The Calculus of Consent*** (with Gordon Tullock), ***Public Finance in Democratic Process***, and ***The Power to Tax*** (with Geoffrey Brennan). He received the Nobel Prize in economics in 1986.

W. ALLAN CRAWFORD received his medical training in Glasgow, Scotland. He joined the Royal Air Force (RAF) and became a specialist in aviation medicine and a research pilot. Upon retirement from the RAF, he emigrated to Australia and there held a post in New South Wales as director of occupational health and director of public health. He has been involved in the public health aspects of lead, ionizing radiation, asbestos, pesticides, photochemical smog, and occupational diseases. He is now a private consultant in environmental and occupational health.

BURT NEUBORNE has been a professor of law at New York University since 1974, where he teaches civil procedure, evidence, federal courts, and constitutional litigation. He served as staff counsel to the New York Civil Liberties Union from 1967 through 1972; as assistant legal director of the American Civil Liberties Union (ACLU) from 1972 through 1974; and as the ACLU's national legal director from 1982 through 1986. He is an active practitioner in the Second Circuit and has argued cases ranging from the legality of the Vietnam War to the constitutionality of restrictions on public utility advertising to the Nassau County 1 percent case. He has written widely in the area of constitutional law. His most recent work is an extended essay on commercial speech, ***Free Speech; Free Markets; Free Choice***, written under the auspices of the Association of National Advertisers.

JODY POWELL, press secretary to Jimmy Carter from 1970 to 1980, is chairman and CEO of Ogilvy and Mather Public Affairs in Washington, D.C. He has been a news analyst for ABC News and wrote a biweekly news column for the Los Angeles Times Syndicate from 1982 to 1987. His book, *The Other Side of the Story,* was published in 1984.

MARK J. REASOR, who received his Ph.D. from Johns Hopkins University, is a professor of pharmacology and toxicology at the West Virginia University Medical Center, Morgantown, West Virginia. Dr. Reasor's research interests lie in the pulmonary toxicity of chemicals. He is a member of the Society of Toxicology and American Society of Pharmacology and

Experimental Therapeutics. Dr. Reasor received the honor of Diplomate of the American Board of Toxicology.

GRAY ROBERTSON is president and founder of ACVA Atlantic Inc., Fairfax, Virginia, a company devoted exclusively to the identification and control of internal pollution problems in public and commercial buildings. ACVA has diagnosed problems and specified solutions in some 30 million square feet of occupied space in the United States, Britain, Japan, Hong Kong, Singapore, Belgium, and Sweden. Born and raised in Liverpool, England, Mr. Robertson graduated from London University. He has worked extensively as a bacteriologist and began focusing on the problem of indoor pollution in 1980.

RENE RONDOU has been the secretary-treasurer of the Bakery, Confectionery, and Tobacco Workers International Union since 1978. He is also vice-president of the Union Label Department at the AFL-CIO and a member of the International Union of Food and Allied Workers Association. He served as president of the Tobacco Workers International Union from 1970 to 1978.

R. EMMETT TYRRELL, JR., is the founder and editor in chief of *The American Spectator*, a political and cultural monthly. He also writes a weekly column that appears in the *Washington Post* and is syndicated by King Features to such papers as the *New York Times, Los Angeles Herald-Examiner*, and *San Francisco Examiner*. Mr. Tyrrell has appeared frequently as a guest on "CBS Morning News," ABC's "Nightline," and NBC's "Summer Sunday USA," among others. He has been the recipient of numerous awards. Tom Wolfe recently described Tyrrell as "the funniest political essayist in years," and in 1979 *Time* magazine named him one of the fifty future leaders of America. Currently he serves as a member of the Academic Advisory Board of the U.S. Naval Academy.

WALTER E. WILLIAMS was born in Philadelphia. He holds a B.A. degree in economics from California State and M.A. and Ph.D. degrees in economics from the University of California at Los Angeles. Dr. Williams has taught at Los Angeles City College, California State University at Los Angeles, and Temple University, and he currently serves on the faculty of George Mason University, Fairfax, Virginia, as John M. Olin Distinguished Professor of Economics. He is the author of over thirty-five publications that have appeared in scholarly journals, has made numerous radio and television appearances, and has served on numerous advisory boards. Dr. Williams was a member of four of President Reagan's advisory and transition teams and has been the recipient of numerous awards. He has participated in numerous debates and conferences in the United States and around the rest of the world. He is frequently called upon to give testimony before both houses of Congress on matters of public policy ranging from labor policy to taxation and spending. His most recent book, ***The State against Blacks*** (1982) has been made into a television documentary, "Good Intentions."

About the Editor

Robert D. Tollison was born and raised in Spartanburg, South Carolina. He attended Wofford College, Spartanburg, where he received his B.A. in 1964 magna cum laude, Phi Beta Kappa. He received an M.A. in economics in 1965 from the University of Alabama, Tuscaloosa, and went on to receive his Ph.D. in economics from the University of Virginia in 1969. Since that time, Dr. Tollison has taught at Cornell, Texas A&M, Virginia Tech, and Clemson University before coming to George Mason University in 1983. He was department head at Texas A&M and is now the director of the Center for Study of Public Choice at George Mason. Dr. Tollison has published over 200 articles in professional economics journals and eleven books. He is a past president of the Southern Economics Association and is a coauthor of a best-selling college textbook, *Economics*. Dr. Tollison has served as a senior staff economist on the President's Council of Economic Advisers and as director of the Bureau of Economics at the Federal Trade Commission. His fields of specialization include public choice, industrial organization, antitrust law and economics, and econometrics.

About the Editor

[illegible]